# Earth Science

**Anne D. Dietz**
**Arthur Troell**
San Antonio Community College

ISBN 978-0-7575-5885-6



Printed in the United States of America
10 9 8 7 6 5 4

# Contents

# Study Tips

1. Come to class ready to learn. Be enthusiastic. Science is fun and interesting–give it a chance!
2. Attend class regularly. There is a direct correlation between attendance and grades. If you aren't attending class regularly, your instructor will assume that you are not interested in passing the class.
3. When you are in class, your attention should be focused on the lecture. Please don't talk to your neighbors. If you have a question, ask your instructor!
4. Please don't leave class unless you have an emergency!
    a. Phone calls, sodas, etc. can wait until after class.
    b. Leaving class disrupts your learning process and that of other class members.
5. Bring your outlines to class.
6. Study the outline and read the assigned pages in the textbook *before* you come to class. Take time to look at the illustrations–they will help you understand the material.
    a. You may find it helpful to break the chapters into sections. Read a section, then stop and review what you have read.
    b. If you find your attention wandering, keep a piece of paper next to your book and put a mark on the page whenever your attention wanders.
7. Take good notes in class.
    a. Write legibly.
    b. Leave space between topics–crowding too many notes on a page can be confusing.
    c. Develop a set of abbreviations that will enable you to write faster.
8. Set aside time to study every day. Don't wait until the night before the exam–there is too much material to learn in one night.
    a. Remember–for every hour you spend in class you should spend at *least* 2 hours studying outside class.
9. Keep track of your grades. If you are unsure about how you are doing in the class, ask your instructor.
10. Ask questions!!! If you are uncomfortable asking questions in class, then ask your instructor after class, during office hours, make a special appointment, etc.

# CHAPTER 1

## Introduction to Earth Science

I. Earth Science Involves the Study of

   A. Physical and Historical Geology

   B. Oceanography

   C. Meteorology

   D. Astronomy

II. Earth Behaves as a System That Has Various Subsystems That Interact with Each Other

   A. Atmosphere

      1. Gases that surround the Earth

      2. Helps protect life from cosmic radiation

   B. Hydrosphere

      1. All water on Earth

   C. Solid Earth (Fig. 1)

      1. Crust

         a. Thin outer layer of the Earth

         b. Oceanic crust–thinner, denser crust composed of dark igneous rock basalt that underlies ocean basins

         c. Continental crust–thicker, less dense "granitic" crust that composes the continents

      2. Mantle–thick, dense layer that underlies the crust

      3. Core

         a. Outer liquid core–iron with some nickel

         b. Solid inner core–iron and nickel

D. Biosphere—all living things on Earth are part of the biosphere

III. Earth's Resources

A. Renewable resources—trees, livestock, wind power, food crops

B. Nonrenewable resources

1. Occur in finite amounts and may take long periods of time for significant amounts to accumulate

2. Oil, gas, coal, metals

IV. Earth's Population

A. In 1830, the population of the Earth was about 1 billion

B. By 1930, the population doubled to 2 billion

C. By 1975, the population doubled to 4 billion

D. In 1998, the population was 6 billion

E. Americans represent only ~6% of the world's population, but use ~30% of world's resources

V. Scientific Investigation

A. Nature behaves in predictable patterns, and scientists seek to discover and understand these patterns

B. Scientists ask a question and then gather data to try and answer the question

C. A hypothesis or hypotheses (untested explanations) are proposed and tested

D. A hypothesis may be accepted, rejected, or modified

E. A hypothesis that has been vigorously tested and not disproven may eventually be accepted as a theory

VI. Plate Tectonics Theory

A. Plate tectonics theory is the unifying theory in geology

1. Helps to explain the occurrence of earthquakes and volcanic eruptions, locations of mountains, etc.

B. Plate tectonics

1. Lithosphere—crust and uppermost brittle mantle (Fig. 1-1)

2. Asthenosphere—upper mantle; capable of slowly flowing (Fig. 1-1)

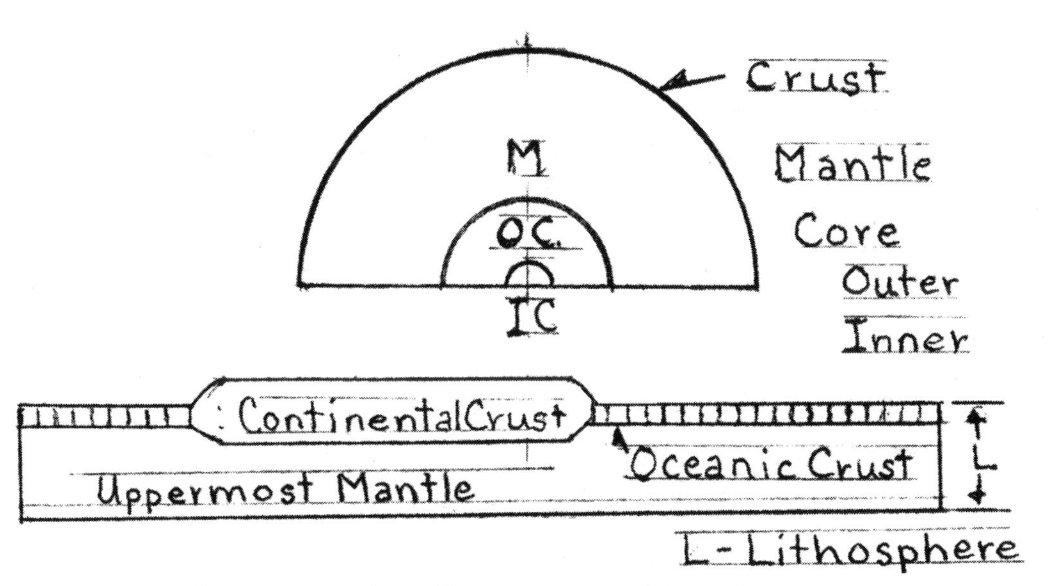


**Figure 1-1** The Different Crusts and the Lithosphere
*© A. Troell, 2008*

3. The lithosphere is broken into plates that move very slowly over the underlying asthenosphere
4. Lithospheric plates interact with each other along their boundaries
5. Types of plate boundaries
   a. Divergent plate boundaries (Fig. 1-2D)
      i. Plates move away from each other
      ii. New oceanic lithosphere is formed
      iii. Mid-Atlantic Ridge that extends down the center of the Atlantic Ocean basin
   b. Convergent plate boundaries
      i. Oceanic-Continental plate boundaries (Fig. 1-2A)
         (a) A plate with oceanic lithosphere on its edge collides with a plate that has a continent on its edge
         (b) Volcanic mountains are formed
         (c) Andes Mountains, South America
      ii. Oceanic-Oceanic plate boundaries (Fig. 1-2B)
         (a) Oceanic lithosphere collides with oceanic lithosphere
         (b) Volcanic islands are formed
         (c) Aleutian Islands (NOT HAWAIIAN ISLANDS!)

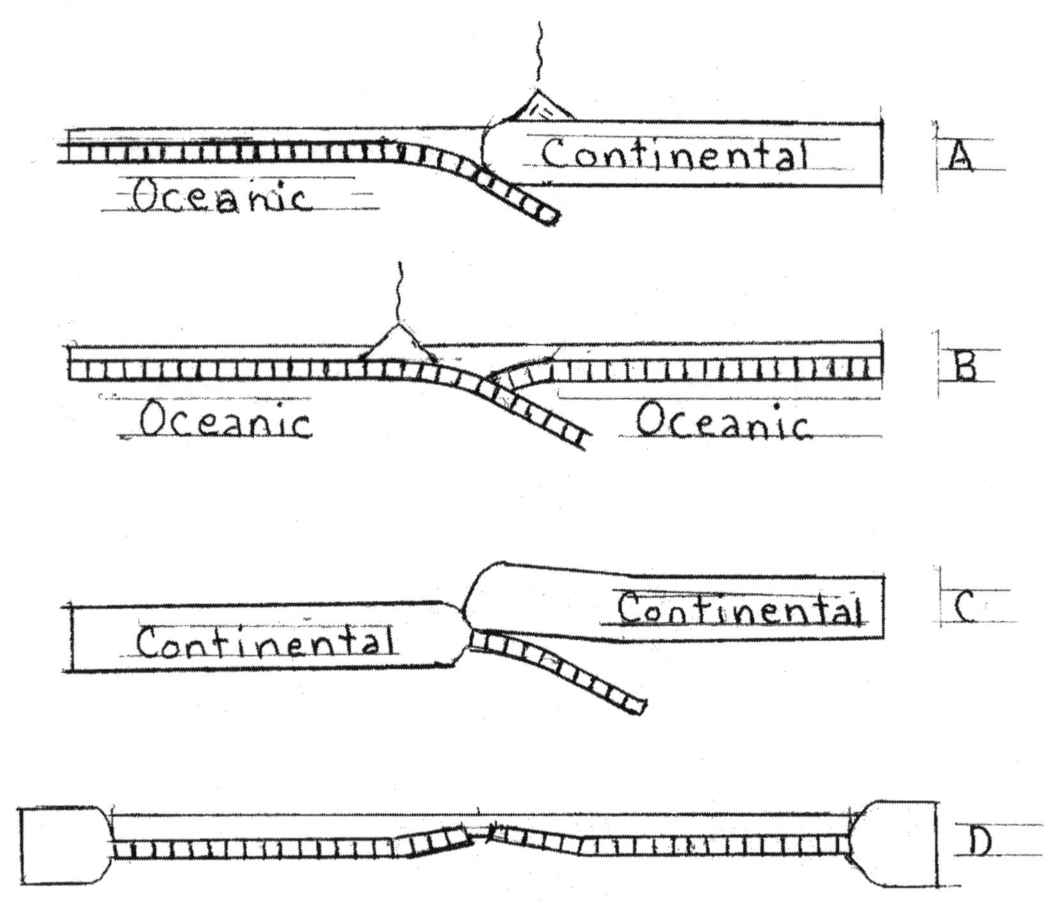


**Figure 1-2** A–C, Plate Convergence; D, Plate Divergence
*© A. Troell, 2008*

iii. Continental-Continental plate boundaries (Fig. 1-2C)
  (a) Continental lithosphere collides with continental lithosphere
  (b) Folded mountains are formed
  (c) Himalayas in Asia

c. Transform plate boundaries
  i. Plates move past each other
  ii. San Andreas Fault in California

VII. Geologic Time

A. Relative time–geologic events are placed in order of occurrence

1. Geologic time scale

a. Units of varying lengths

b. Eons, Eras, Periods, Epochs

B. Absolute time–time in years

1. Radiometric dating

a. Half-life of radioactive decay

b. Examples

i. Long-lived isotope pairs

ii. Carbon-14

2. Tree-ring dating

3. 4.5 billion years of earth history

a. Phanerozoic Eon

i. Cenozoic Era (66 my–0)

ii. Mesozoic Eon (245–66 my)

iii. Paleozoic Eon (545–245 my)

b. Precambrian (4.5 by–545 my)

# NOTES

# CHAPTER 2

## Minerals

I. Rocks Are Composed of Minerals

II. Minerals = Building Blocks of Rocks

   A. Characteristics

      1. Naturally occurring

      2. Inorganic = Never living

      3. Solid, never liquids or gases

      4. Narrowly defined chemical composition

      5. Atoms are arranged in an orderly repeating pattern

III. Atoms

   A. Elements are composed of atoms

   B. Parts of an atom

      1. Nucleus

         a. Protons–positive charge

         b. Neutrons–no charge

      2. Electrons–negative charge

         a. Orbit the nucleus in energy levels or shells

      3. Atomic number

         a. Number of protons in the nucleus

         b. Example

            i. Carbon is number 6 on the periodic table

            ii. Atoms of carbon always have 6 protons

Free atom = Neutral, No Charge

IV. Bonding

A. Compounds—two or more elements bonded together in definite proportions

B. Bonding involves the electrons in the outermost energy level

C. Types of bonds

1. Ionic bonding

a. Atoms gain or lose one or more electrons

b. Example—Halite (NaCl)—Fig. 2-1

2. Covalent bonding

a. Atoms share electrons

b. Example—Diamond (Fig. 2-2) / Graphite = Bonded in sheets (lead pencil)

V. Mass Numbers and Isotopes

A. Mass number

1. Sum of protons and neutrons in the nucleus

2. Isotopes

a. Varieties of same element that have different numbers of neutrons

b. Example

i. $C^{12} = 6$ protons, 6 neutrons

ii. $C^{14} = 6$ protons, 8 neutrons

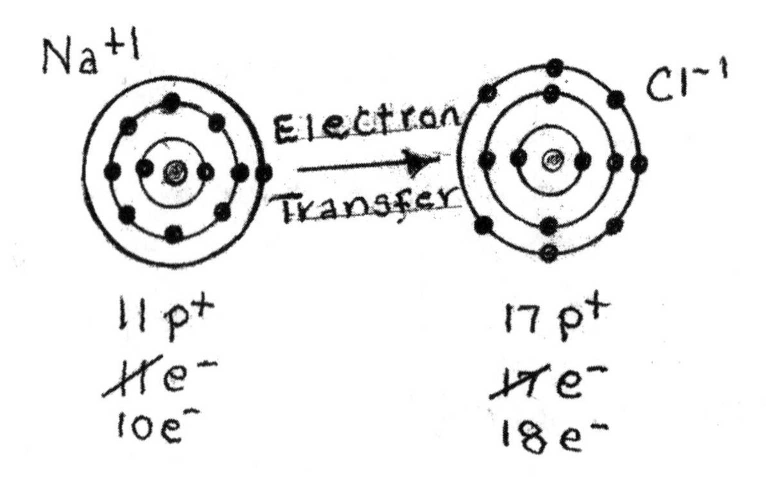


Halite (NaCl)

**Figure 2-1** Ionic Bonding

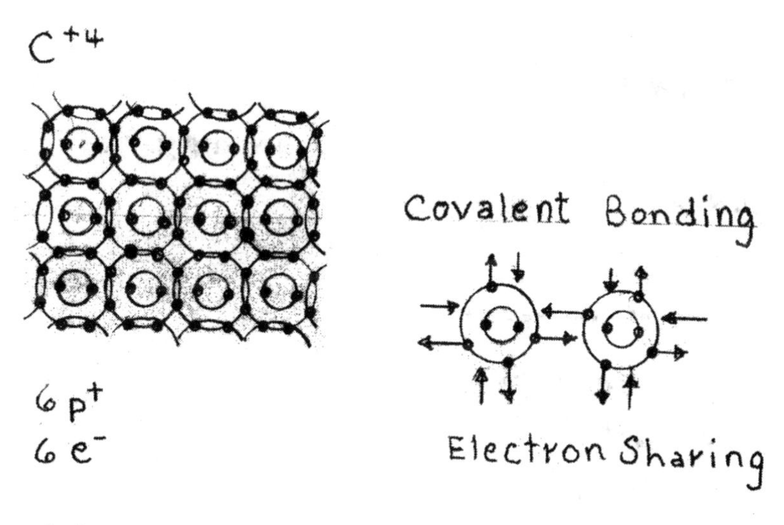


**Figure 2-2** Diamond–Covalent Bonding and Electron Sharing
*© A. Troell, 2008*

VI. Properties of Minerals

A. There are about 3,500 known minerals

1. Most minerals are compounds–composed of two or more elements bonded together

B. Minerals have different properties due to the elements present and the way the elements are bonded together

C. Crystal shape

1. The external shape of a mineral crystal is a reflection of the arrangement of the elements
2. Crystals form when there is enough space for well-shaped crystals to form
3. Crystal faces–flat surfaces
4. Quartz–crystals are six-sided pillars with pyramids on the ends (Fig. 2-3)
5. Halite–cubic crystals (Fig. 2-4)
6. Calcite–rhombohedral crystals (Fig. 2-5)

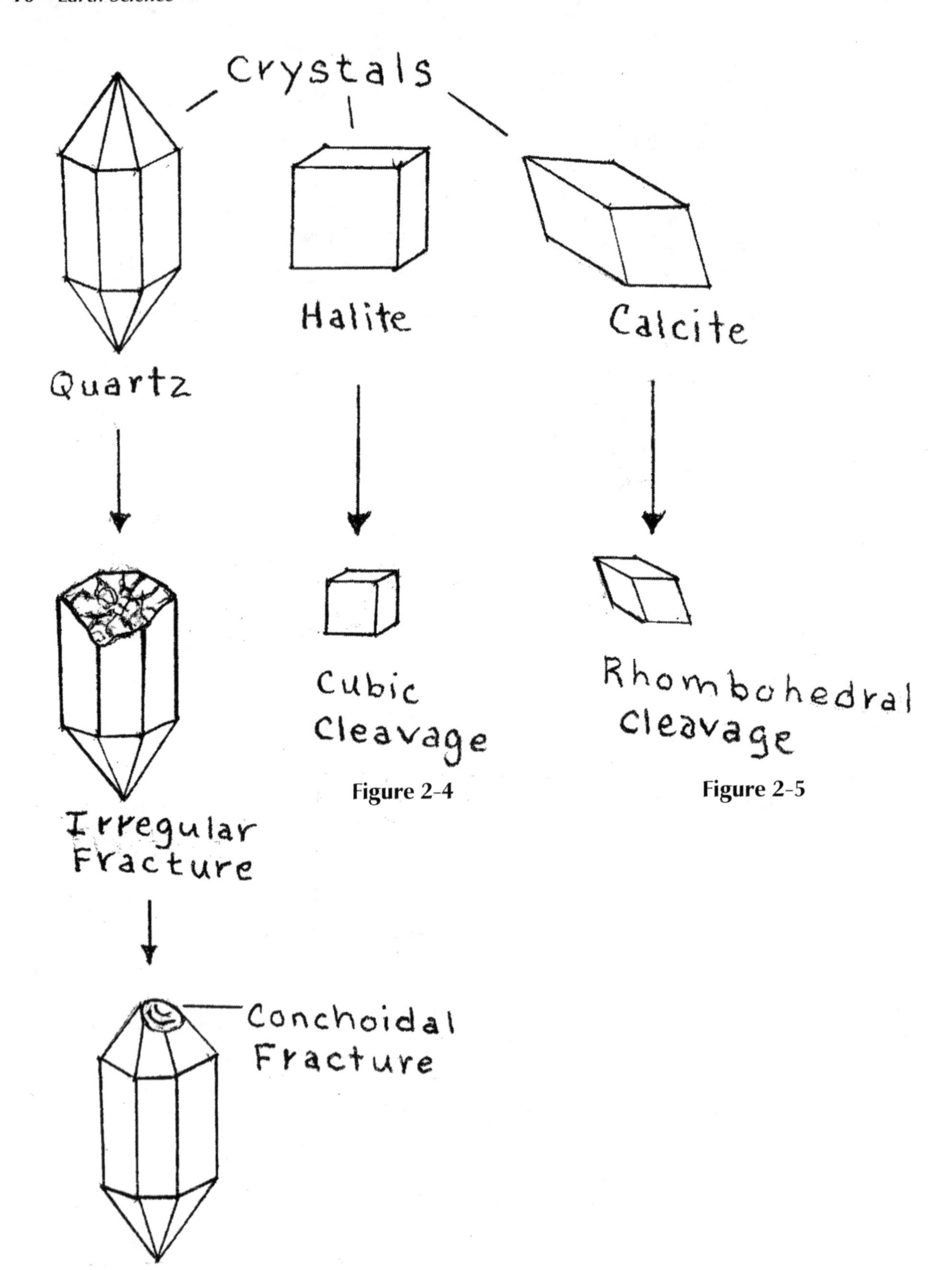


Figure 2-4

Figure 2-5

Figure 2-3

D. Luster

1. Appearance or quality of light reflected from the surface of a mineral
2. Types of luster
    a. Metallic luster
        i. Only a limited number of minerals have metallic luster
    b. Nonmetallic luster
        i. A large number of minerals have a nonmetallic luster, so there are different subcategories such as vitreous (glassy) and earthy (dull)

E. Color

1. Some minerals are color specific
2. Many minerals, such as quartz, occur in a variety of colors

F. Streak

1. Color of a mineral when powdered
2. Use a streak plate

G. Hardness

1. How resistant a mineral is to abrasion
2. Mohs hardness scale
    a. Scale of relative hardness
    b. Talc–1, diamond–10
3. Use objects of known hardness to assess hardness of minerals
    a. Fingernail–2, Copper penny–3.5, Glass–5.5
4. NOTE–ALWAYS PUT THE COPPER PENNY AND GLASS ON A SOLID SURFACE WHEN TESTING HARDNESS!

H. Cleavage

1. Tendency of mineral to break along weak bonds in the atomic structure
    a. Produces flat planes on the mineral
2. Count the number of directions of cleavage (look for very flat planes) and the angle between the cleavage planes
    a. One direction of cleavage
    b. Two directions of cleavage

c. Three directions of cleavage at right angles–halite (Fig. 2-4)

d. Three directions of cleavage not at right angles–calcite (Fig. 2-5)

e. Four directions of cleavage

f. Six directions of cleavage

I. Fracture

1. Some minerals break unevenly (do not exhibit cleavage)

2. Conchoidal fracture

a. Quartz breaks along smooth, curved surfaces (Fig. 2-3)

J. Specific Gravity

1. The ratio of the weight of a mineral to the weight of an equal volume of water

2. Most minerals have SG = 2.5–3.5

3. Some minerals have a high specific gravity and feel heavy–Galena (PbS)

K. Specific properties

1. Calcite–effervesces in HCl

2. Talc–feels soapy

3. Halite–tastes salty

4. Magnetite–magnetic

L. Note: Always wash your hands after handling minerals in class!

VII. Mineral Groups

A. Most abundant elements in the Earth's crust

1. Oxygen (O), silicon (Si), aluminum (Al), iron (Fe), calcium (Ca), sodium (Na), potassium (K), magnesium (Mg)

B. Minerals are grouped according to the elements they are composed of

C. Silicate Group

1. Most abundant group of minerals

2. Minerals in this group all have the silicon-oxygen tetrahedron $(SiO_4)^{-4}$, which is a complex ion

3. Silicate minerals are constructed of silicon-oxygen tetrahedral joined together in silicate structures
    a. Most tetrahedra are joined by sharing oxygen atoms with neighboring tetrahedra (except single tetrahedron)
4. Silicates structures
    a. Isolated (single) tetrahedron
        i. Linked to neighboring tetrahedral by positive ions
    b. Single chains
    c. Double chains
    d. Sheets–micas (muscovite, biotite)
    e. Three-dimensional networks–Quartz ($SiO_2$)

D. Carbonate Group
1. Minerals in this group all have $(CO_3)^{-2}$ complex ion
2. Calcite–$CaCO_3$

E. Halide Group
1. Minerals in this group all have halide elements such as chlorine (Cl-1), fluorine ($F^{-1}$)
2. Halite–NaCl

F. Oxide Group
1. Minerals in this group have oxygen ($O^{-2}$) in their atomic structure
2. Hematite–$Fe_2O_3$

G. Sulfates $(SO_4)^{-2}$
1. Gypsum–$CaSO_4\ 2H_2O$
2. Used to make wallboard

H. Native elements
1. Minerals that are composed of only one element
2. Native copper, silver and gold, diamond, graphite

# NOTES

# CHAPTER 3

## Igneous Rocks

I. Rock Cycle

   A. There are three types of rocks–igneous, sedimentary, and metamorphic

   B. The rock cycle shows the relationships between the three types of rocks and how one type of rock may give rise to a different type of rock

II. Igneous Rocks–"Fire-Formed"

   A. Igneous rocks form from magmas or lavas as they cool and minerals crystallize (Fig. 3-1)

      1. Magma–molten rock within the earth

         a. Intrusive (or plutonic) igneous rocks form from magmas

      2. Lava–molten rock that erupts on the earth's surface

         a. Extrusive (or volcanic) igneous rocks form from lavas

      3. Both magma and lava may be composed of solids, liquids, and gases

| Classification of Igneous Rocks | | | | | |
|---|---|---|---|---|---|
| | **Felsic** | **Intermediate** | **Mafic** | **Ultra-mafic** | |
| **Color** | **Pink** | **Gray** | **Black** | **Black & Green** | |
| Minerals | Quartz ← | Feldspar | → | Olivine | |
| Fine Crystals | Rhyolite | Andesite | Basalt | | Extrusive (volcanic) |
| Coarse Crystals | Granite | Diorite | Periodotite | | Intrusive (Plutonic) |

**Figure 3-1** Classification of Igneous Rocks

| Textures | |
|---|---|
| Pyroclastic | Ash (Tuff) |
| Glassy | Obsidian |
| Vesicular | Scoria (coarse, Pumice (Fine) |
| Crystalline | Fine (microscopic), Coarse (macro-) |

**Figure 3-2** Textures
*© A. Troell, 2008*

4. As magmas and lavas cool, minerals crystallize and grow until they meet the edges of other minerals, so igneous rocks are composed of interlocking minerals
5. Igneous rocks are classified according to their texture and composition

III. Igneous Rock Textures—Size, Shape, and Arrangement of Minerals (Fig. 3-2)

A. Coarse-grained (Phaneritic) texture

1. All of the minerals are about the same size and are visible to the naked eye
2. Phaneritic texture develops because magma cools slowly
   a. The surrounding rocks don't conduct heat well, so magmas may stay hot for long periods of time thus giving minerals time to grow larger

B. Fine-grained (Aphanitic) texture

1. All of the minerals are about the same size, but are too small to see with the naked eye
   a. Aphanitic texture develops when lavas cool quickly
      i. Many small minerals develop, but lava cools so quickly there is no time for minerals to grow large

C. Porphyritic texture

1. Minerals are of two sizes due to different rates of cooling
   a. Phaneritic-porphyritic texture
      i. All the minerals are visible, but some minerals are significantly larger

b. Aphanitic-porphyritic texture
i. Larger minerals in a matrix of smaller minerals
ii. Slow rate of cooling, then fast rate of cooling
(a) Some minerals formed from magma as it slowly cooled, then the magma erupted onto the surface and cooled quickly

D. Glassy texture
1. Lava cools instantly, so no minerals are present
2. Obsidian ("volcanic glass")
3. Pumice–light-colored
4. Scoria–dark-colored

E. Pyroclastic texture
1. Pyroclastic material forms when material is ejected explosively during a volcanic eruption
2. Range from dust and ash (very small) to lapilli and bombs
3. Pyroclastic materials consolidate to form igneous rocks with pyroclastic textures

IV. Igneous Rock Compositions
A. Bowen's Reaction Series
1. Bowen experimented with how minerals crystallize from a magma
2. As magma cools, minerals crystallize
3. Different minerals crystallize at different temperatures
4. Ideally, when a magma cools, the first mineral to crystallize is olivine; the last mineral to crystallize is quartz
5. Minerals that crystallize at about the same temperature are found together in the same igneous rock

V. Classification of Igneous Rocks (Fig. 3-3)
A. Felsic (granitic) composition
1. Rocks that are composed mostly of light-colored minerals such as quartz and feldspar

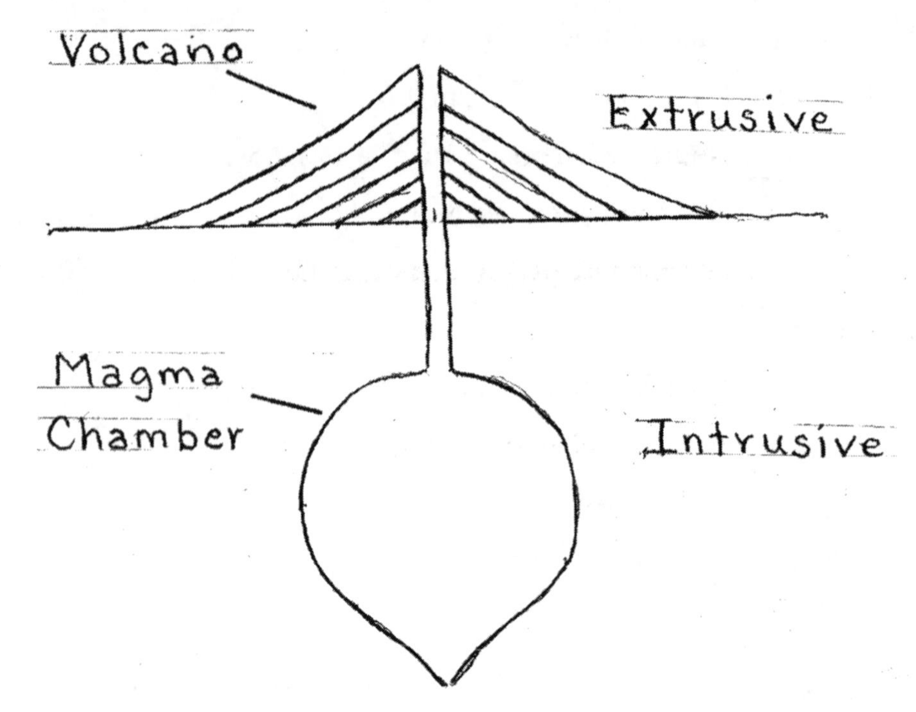


**Figure 3-3** Volcano, Magma Changer, etc.
*© A. Troell, 2008*

2. Granite
    a. Coarse-grained, light-colored (pink to white)
    b. Intrusive equivalent of rhyolite
    c. Often compose the cores or roots of mountains
3. Rhyolite
    a. Fine-grained, light-colored (pink to white)
    b. Extrusive equivalent of granite

B. Intermediate composition
1. Rocks that are composed of light- and dark-colored minerals
2. Diorite
    a. Coarse-grained, often black and white in color
    b. Intrusive equivalent of andesite
3. Andesite
    a. Fine-grained, often gray or green in color
    b. Extrusive equivalent of diorite
    c. Forms in association with volcanic activity at convergent plate boundaries

C. Mafic (basaltic) composition

1. Gabbro

    a. Coarse-grained, usually black

    b. Intrusive equivalent of basalt

    c. Composes lower part of oceanic crust

2. Basalt

    a. Fine-grained, usually black

    b. Extrusive equivalent of gabbro

    c. Composes upper part of oceanic crust

D. Ultramafic composition—composed mostly of minerals high in Fe & Mg (such as olivine)

1. Peridotite

    a. Coarse-grained, often green in color

    b. Composes Earth's mantle

# NOTES

# CHAPTER 4

## Weathering

I. Weathering
   A. Physical and chemical breakdown of rocks at/near Earth's surface
   B. Weathering occurs because the minerals that compose rocks aren't stable at conditions found at Earth's surface
      1. As granite is exposed at the surface, feldspar minerals aren't stable, so they begin to break down and become more stable under the lower temperatures and pressures found at Earth's surface
   C. Mechanical and chemical weathering work together at Earth's surface
      1. Mechanical weathering increases the surface area for chemical weathering to attack
   D. Erosion
      1. Removal and transportation of material

II. Mechanical Weathering
   A. Frost Wedging (Fig. 4-1)
      1. Alternate freezing/thawing of water in fractures in rocks causes rocks to break down over time
      2. Common in mountainous areas
      3. May create talus slopes
         a. Cone-shaped deposits of loose material at the base of a slope
         b. When talus lithifies, it may form breccia
   B. Unloading (Pressure Release)–(Fig. 4-2)
      1. As rocks (especially granites) are exposed at the surface, the pressure is relieved

**Figure 4-1** Frost Wedging and Talus Slope
*© A. Troell, 2008*

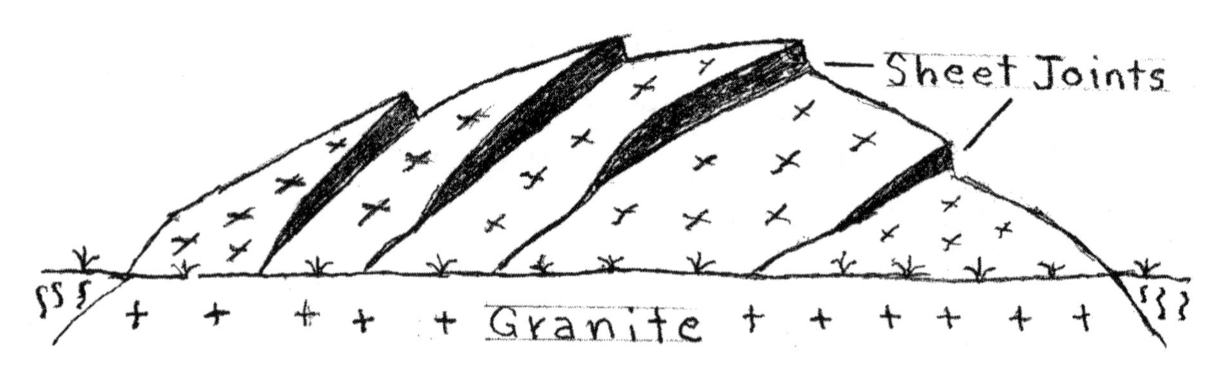


**Figure 4-2** Exfoliation Dome
*© A. Troell, 2008*

2. The outer surface of the granite fractures along sheet joints, which are parallel to the surface
3. Over time, pieces of rocks break off along the sheet joints
4. Unloading creates exfoliation domes
    a. Enchanted Rock, TX
    b. Stone Mountain, GA

C. Activity of Animals and Plants
1. Burrowing animals may help break down rocks
2. Root wedging
    a. Plant roots grow into fractures in rocks and over time may cause rocks to break down

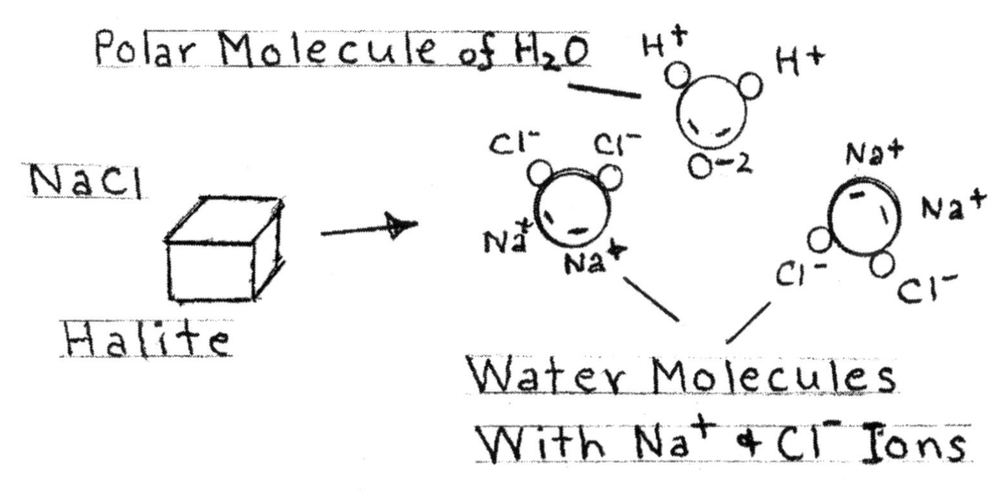


**Figure 4-3** Polar Molecule of $H_2O$
*© A. Troell, 2008*

III. Chemical Weathering

A. Minerals are altered by the addition and/or removal of ions

B. Water is very important in chemical weathering

C. Types of chemical weathering

1. Dissolution (Fig. 4-3)

   a. Halite dissolves in water

   b. Rainwater + $CO_2$ produces $H_2CO_3$ (carbonic acid)

      i. Carbonic acid dissolves calcite

2. Oxidation

   a. Iron in water reacts with oxygen to form hematite

3. Hydrolysis

   a. H+ in carbonic acid replaces elements such as Al, K, Na in feldspars in granites

   b. Chemical weathering of feldspars produces clay minerals

   c. Al, Na, K ions go into solution, as well as some $SiO_2$

   d. Quartz grains, which are resistant to chemical weathering, are released and transported, then deposited to become part of a new rock

# NOTES

# CHAPTER 5

## Sedimentary Rocks

I. Sedimentary Rocks "Settling"

A. Sedimentary rocks form from the products of chemical and mechanical weathering (Fig. 5-1)

B. Detrital Sediment = solids

1. Particles derived primarily from the products of mechanical weathering

C. Chemical Sediment

1. Ions and compounds in solution that derived from chemical weathering of rocks

2. These ions and compounds eventually crystallize from solution to become solid particles

D. Erosion

1. Removal and transportation of sediment by running water, wind, glaciers, or currents

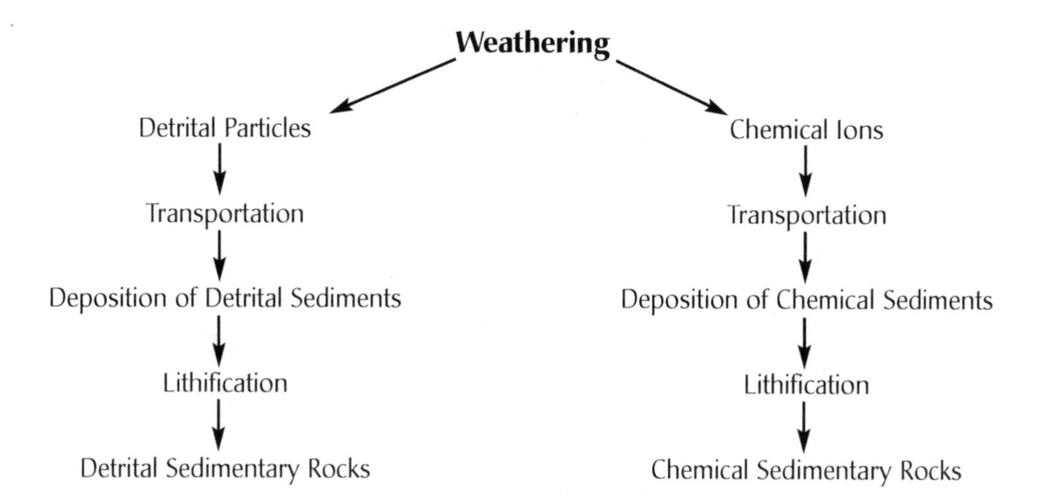


**Figure 5-1** Weathering
*© A. Troell, 2008*

- E. When the transporting agent loses energy, sediment is deposited
- F. Importance of sedimentary rocks
    1. Oil and gas are trapped in sedimentary rocks
    2. Two types of coal are classified as sedimentary rocks
    3. Sedimentary rocks preserve clues about ancient environments of deposition
    4. Fossils are commonly preserved in sedimentary rocks

II. Lithification

- A. The process of turning sediment into sedimentary rock
- B. Lithification may include
    1. Compaction
        - a. Weight of overlying sediment causes grains to compact more closely together
    2. Cementation
        - a. Minerals precipitate from solution in the pores spaces (openings) between grains
        - b. Types of cements
            - i. Quartz
            - ii. Calcite
            - iii. Iron oxide

III. Detrital Sedimentary Rocks

- A. Form from the lithification of detrital sediments
- B. Classified primarily by grain size
- C. Sorting
    1. Well-sorted
        - a. Grains that compose the rock are all about the same size
        - b. Running water and wind
    2. Poorly-sorted
        - a. Grains that compose the rock are different sizes
        - b. Glaciers

D. Degree of rounding
   1. Angular grains
      a. Indicate a short distance of transport
   2. Well-rounded particles
      a. Indicate a long distance of transport

E. Common minerals that compose detrital sedimentary rocks
   1. Quartz
   2. Feldspar
   3. Clay minerals (derived from chemical weathering of feldspar minerals)

F. Classification of Detrital Sedimentary Rocks (Fig. 5-2)
   1. Conglomerate
      a. Composed of rounded, gravel-sized grains
      b. May be found where energy is high—river beds
   2. Breccia—composed of angular, gravel-sized grains
   3. Quartz Sandstone
      a. Composed of >90% quartz grains
      b. May form from beach or sand dune deposits
   4. Arkose
      a. Sandstone composed of >25% feldspar grains
      b. May form near granites that are weathering

| Classification of Detrital Sedimentary Rocks | | | |
|---|---|---|---|
| **Partical Size** | **Sediment Name** | **Lithification** | **Rock Name** |
| >2 mm | Gravel | Cementation | Conglomerate* or Breccia |
| $\frac{1}{16}$–2 mm | Sand | Cementation | Sandstone Quartz, Arkose |
| <$\frac{1}{16}$ mm | Mud | Compaction | Shale |
| *Rock Name Is Breccia If Particles are Angular | | | |

**Figure 5-2** Classification of Detrital Sedimentary Rocks

5. Shale
    a. Composed of grains that are too small to see without magnification
    b. Fissile–splits along flat planes that reflect arrangement of platy clay-sized minerals

IV. Chemical Sedimentary Rocks
  A. Rocks formed as a result of lithification of chemical sediments
  B. Classified according to composition and grain size
  C. Classification of Chemical Sedimentary Rocks (Fig. 5-3)
    1. Limestones–composed of calcium carbonate
        a. Travertine–limestone that is deposited in caves
        b. Coquina–limestone composed of visible shells and shell fragments that have been cemented together
        c. Chalk–limestone composed of microscopic calcareous ($CaCO_3$) hard parts of marine plants and animals

**Classification of Chemical Sedimentary Rocks**

| Texture | Composition | Rock Name | |
|---|---|---|---|
| Particulate | Calcite | Coquina | ↑ |
| Particulate | Calcite | Chalk | Limestone |
| Particulate | Calcite | Lime Mudstone | |
| Crystalline | Calcite | Travertine | ↓ |
| Crystalline | Dolomite | Dolostone | |
| Crystalline | Quartz | Chert | |
| Crystalline | Gypsum | *Rock Gypsum | |
| Crystalline | Halite | *Rock Salt | |
| Particulate | Organic Matter | Coal | |

*Evaporites

**Figure 5-3** Classification of Chemical Sedimentary Rocks
*© A. Troell, 2008*

2. Chert–composed of the microscopic siliceous ($SiO_2$) hard parts of marine plants and animals
3. Evaporites
   a. Composed of minerals that precipitate when bodies of water evaporate
   b. Rock salt–composed of halite
   c. Rock gypsum–composed of gypsum
4. Coal (Fig. 5-4)
   a. Composed of altered plant material
   b. Coal forms in swampy environments where there is little/no oxygen in the water
   c. Plant material only partially decays and goes through stages of alteration
   d. Peat
      i. Partially altered plant material
      ii. Essentially the sediment stage of coal
   e. Lignite Coal
      i. When peat undergoes burial and compaction, it may alter to lignite coal
      ii. Soft brown coal
   f. Bituminous Coal
      i. If lignite coal undergoes further burial and compaction it may form bituminous coal
      ii. Soft black coal

**Stages in Formation of Coal**

→ Increasing Depth of Burial →

Peat → Lignite → Bituminous Coal → Anthracite

Peat ↑ Sediment, (Swamp Deposit)

Lignite: Soft, Brown

Bituminous Coal: Soft, Black

↑ Metamorphism

Anthracite: Hard, Black

**Figure 5-4** Stages in Formation of Coal
*© A. Troell, 2008*

g. Anthracite Coal

i. If bituminous coal undergoes metamorphism, it will alter to anthracite coal

ii. Hard black coal

V. Sedimentary Structures

A. Feature preserved in sedimentary rocks

B. Sedimentary rocks are deposited in layers or beds

C. Beds are separated by bedding planes

1. Flat planes that indicate the end of one episode of deposition and the beginning of another period of deposition

D. Ripple marks

1. Ridges on the surface of sedimentary rocks

2. Wave-formed ripples

a. Formed by back and forth movement of waves

b. Ripples are symmetrical in shape

3. Current ripples

a. Formed by currents in rivers or oceans or by wind moving in one direction

b. Ripples are asymmetrical in shape

E. Cross-beds

1. Inclined layers of sand within a larger bed of rock

2. Sand dunes often exhibit cross-beds formed by changing wind directions

F. Mud cracks—Form when clay-rich sediment dries and shrinks

# NOTES

# NOTES

# CHAPTER 6

## Metamorphic Rocks

I. Metamorphic "Change Form"

   A. Preexisting igneous, sedimentary, or metamorphic rocks are altered in the solid state through the effects of heat, pressure, and/or chemically active fluids

   B. Metamorphic change occurs because minerals in a preexisting rock are out of equilibrium when they are exposed to new conditions

   C. Changes that may occur

      1. Minerals may become aligned (foliated texture)

      2. Minerals may grow larger

      3. New minerals may form

   D. Metamorphic grade

      1. Low-grade metamorphism

         a. Occurs at low temperatures, low pressure

      2. High-grade metamorphism

         a. Occurs at high temperatures, high pressures

   E. Metamorphic Settings

      1. Contact metamorphism (Fig. 6-1)

         a. Preexisting rocks are affected by the heat from nearby magmas

      2. Regional metamorphism (Fig. 6-2)

         a. Occurs during mountain building at plate boundaries

            i. Rocks are affected by high temperatures and pressures

         b. Usually affects a large area

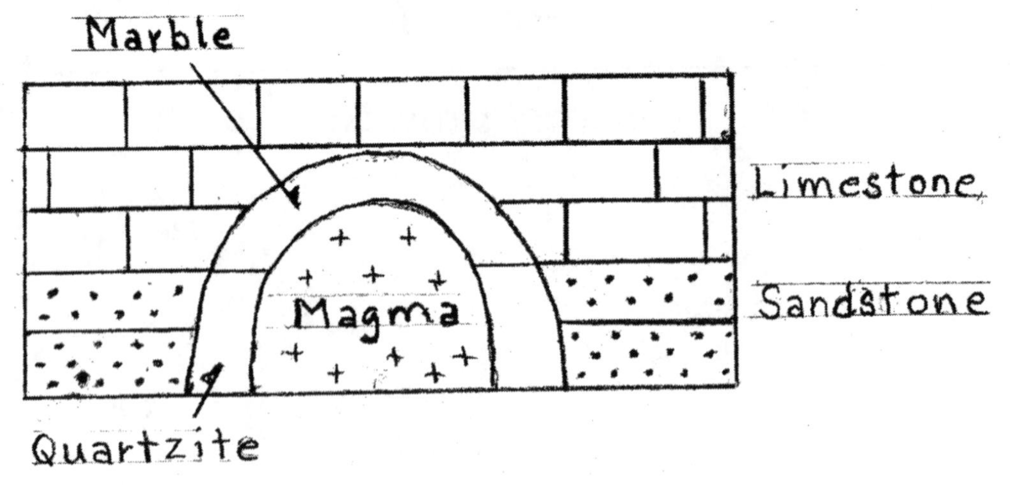


**Figure 6-1** Contact Metamorphism
*© After A. Dietz*

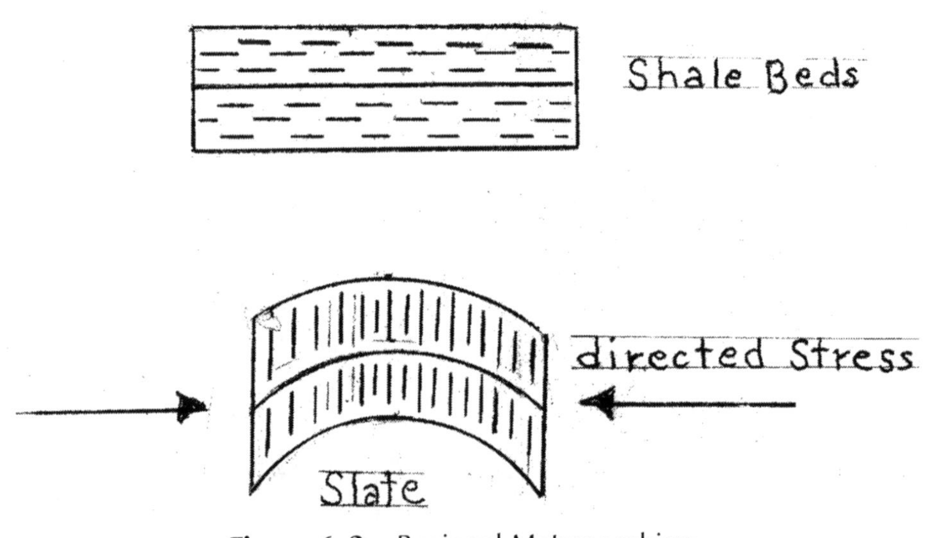


**Figure 6-2** Regional Metamorphism
*© After A. Dietz*

II. Agents of Metamorphism

A. Heat

1. Heat provides energy to drive chemical reactions

2. Sources of heat

a. Geothermal gradient

i. Increasing temperature with increasing depth in Earth

b. Nearby magma

i. Limestone may be altered to marble (Fig. 6-1)

B. Pressure

1. Confining pressure (lithostatic pressure)

    a. Pressure is the same in all directions

    b. Minerals may recrystallize as smaller, denser minerals

2. Directed stress (Fig. 6-2)

    a. Stress is greater in one direction

    b. Causes foliation (alignment of minerals)

    c. Occurs at convergent plate boundaries

C. Chemically active fluids

1. Hot water with ions

2. New minerals may form

III. Classification of Metamorphic Rocks (Fig. 6-3)

A. Foliated metamorphic rocks

1. Exhibit an alignment of minerals

2. Usually form during regional metamorphism

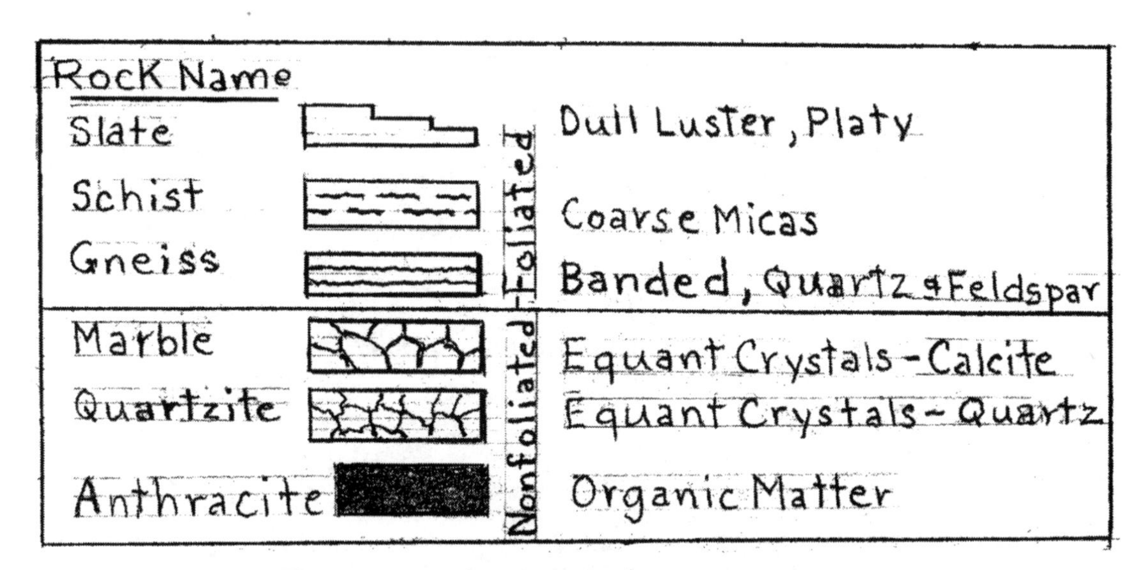


**Figure 6-3** Classification of Metamorphic Rocks

B. Metamorphic rocks with foliated textures
   1. Slate
      a. Fine-grained
      b. Slaty cleavage
         i. Splits along flat planes
         ii. Reflects alignment of clay-sized minerals
      c. Slate forms through the low-grade metamorphism of shale
   2. Schist
      a. Coarse-grained rocks composed of >50% platy minerals
      b. Wavy foliation
      c. Intermediate- to high-grade metamorphism
   3. Gneiss
      a. Distinctive light and dark bands
      b. Coarse-grained
      c. High grade metamorphism

C. Nonfoliated metamorphic rocks
   1. Lack platy minerals
   2. Rocks are often composed of one primary mineral

D. Metamorphic rocks with nonfoliated textures
   1. Quartzite
      a. Composed primarily of quartz grains
      b. Parent rock—quartz sandstone
         i. Quartz grains have been fused together
      c. Grainy texture
      d. May form during regional or contact metamorphism
   2. Marble
      a. Composed primarily of calcite grains
      b. Parent rock—limestone
         i. Calcite grains recrystallize and grow larger
      c. Coarsely crystalline
      d. May form during regional or contact metamorphism

3. Anthracite coal ← like Galena – No weight/very light
   a. Hard, black coal
   b. Parent rock—bituminous coal ← Regional Metamor

NOTE: Figure 6-4 is a modified rock cycle that includes all the rocks we have studied in class.

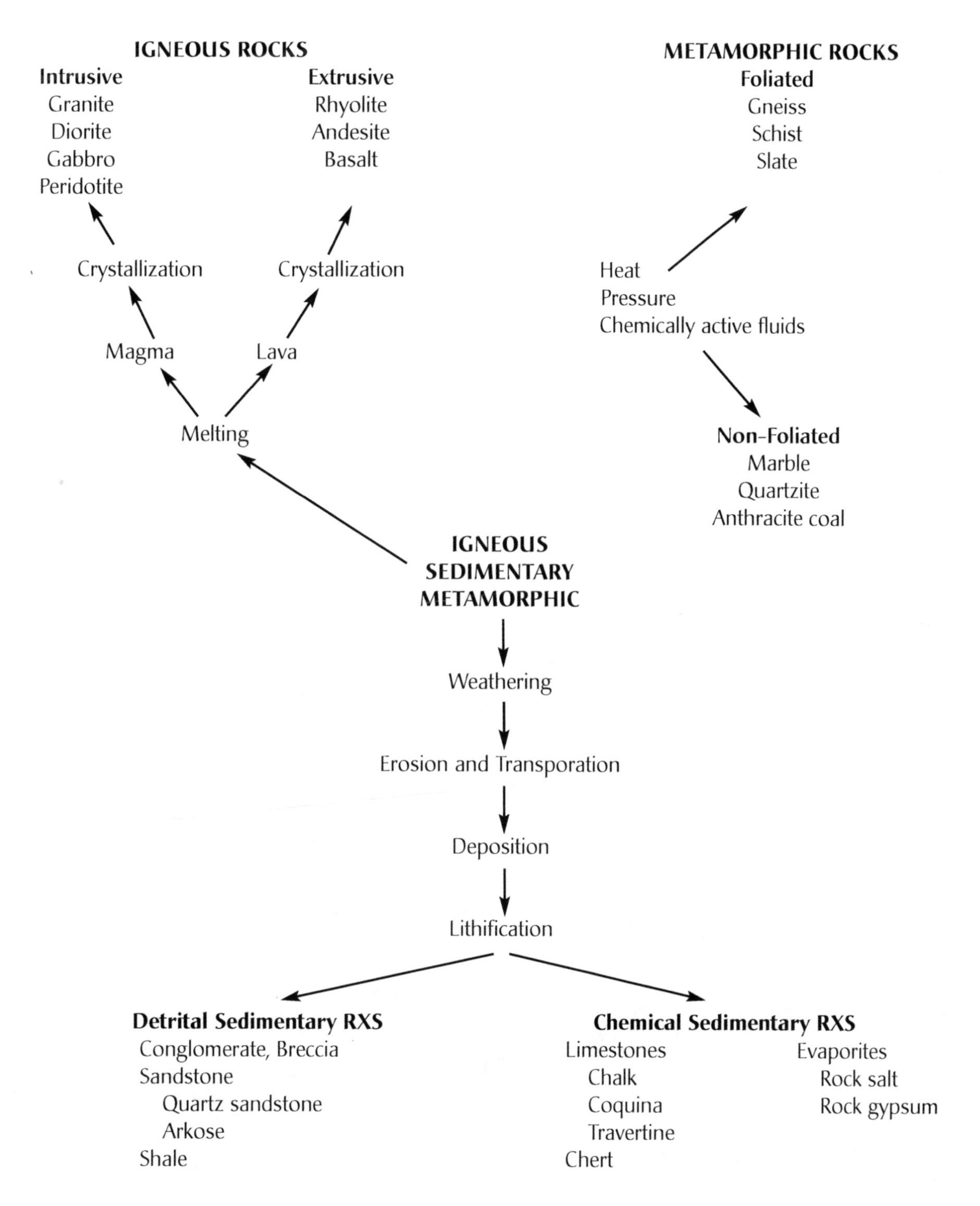


**Figure 6-4** Sedimentary Rocks

# NOTES

# CHAPTER 7

## Soils

I. Soil–Mixture of mineral and organic matter, water, and air that is capable of supporting the growth of plants

II. Origin of Soil–Due to weathering of earth materials (rocks and minerals)

   A. Residual soils–formed by the chemical weathering of rocks in-place

   B. Transported soils–formed by the chemical weathering of sediments (volcanic, windblown, or river sediment)

III. Climates and Soils

   A. Polar climates

      1. Thin soils

      2. Chemical weathering is minor

   B. Midlatitude climate

      1. Desert climates

         a. Rainfall less than 10 inches rain/year (average)

         b. Thin, poorly developed soils

         c. Mechanical weathering dominant

      2. Dry temperate climates

         a. Rainfall 10–25 inches/year (average)

         b. Thin soil layers

      3. Wet temperate climates

         a. Rainfall 25–70 inches rain/year (average)

         b. Thick soil layers

   C. Wet tropical climates

      1. Rainfall greater than 80 inches/year (average)

      2. Very thick soils

      3. Hot weather throughout the year

IV. Soil Horizons (layers)

A. Basic soil horizons (Fig. 7-1)

1. O horizon

    a. Leaves, twigs, and stems at top

    b. Humus below

2. A horizon (topsoil)

    a. Zone of leaching and eluviation

        i. Leaching—removal of soluble material

        ii. Eluviation—removal of fine particles

    b. High biological activity

    c. Contains humus (dark-colored)

3. B horizon (subsoil)

    a. Zone of material leached and eluviated from the A horizon, especially

        i. Clays

        ii. Iron

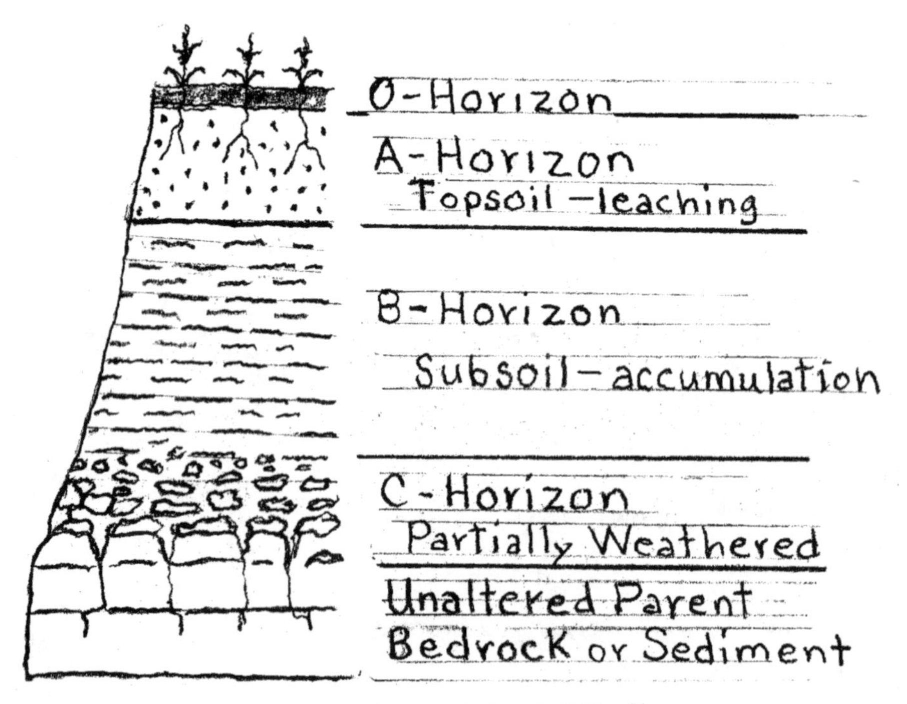


**Figure 7-1** Soil Profile
© A. Troell, 2008

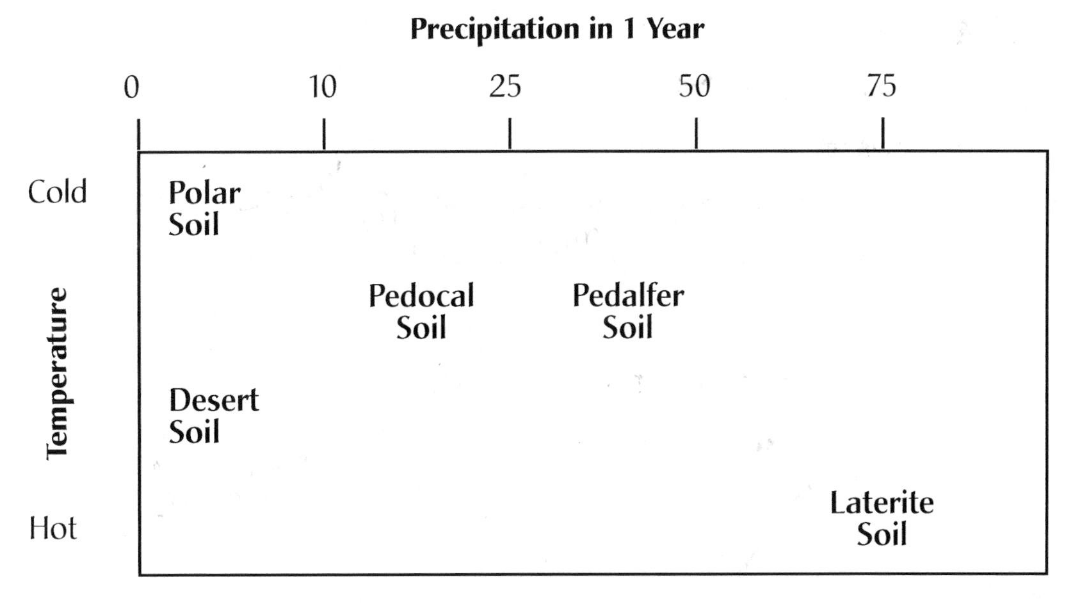


**Figure 7-2** Climate and Soil Type
*© A. Troell, 2008*

4. C horizon
    a. Partially altered parent material
    b. May be bedrock or sediment

V. Climate and Soil Type (Fig. 7-2)

A. Polar soils
1. Soils beneath tundra
2. Thin soils
    a. Too cold for chemical weathering

B. Desert soils
1. Thin and patchy
2. Contain alkali–sodium and potassium carbonates

C. Pedocal soils
1. Thin A and B horizons
2. Has K horizon at the base of the B horizon
    a. Caliche may be deposited here
        i. White crusty calcium carbonate deposited when water evaporates

D. Pedalfer soils

1. Thick A and B horizons
2. B horizon contains abundant iron oxides and hydroxides
3. E horizon (white leached zone) developed at base of A horizon beneath forests
4. Found in eastern part of United States and Pacific Northwest

E. Laterite soils

1. Thick red soils of wet tropics
2. Contain very little humus
3. Consists entirely of iron oxides, hydroxides, and aluminum hydroxides (bauxite)
   a. Bauxite–principle ore of aluminum
4. Clays and even quartz leached away
5. Laterite soils are found beneath rain forests

VI. Expansive Soils

A. Contain clay minerals that expand when wet

B. May cause foundation problems for homes

C. Occur in Bexar County

# NOTES

# NOTES

# CHAPTER 8

## Mass Wasting

I. Mass Wasting

   A. Occurs when earth materials move down a slope under the influence of gravity

   B. Mass wasting occurs naturally or may be caused by human activity

II. Causes of Mass Wasting

   A. Addition of water may cause mass wasting because

      1.' Water allows particles to move over each other

      2. Water adds weight to material

   B. Oversteepening of slopes

      1. Angle of Repose

         a. The steepest angle at which unconsolidated material remains stable

            i. Sand–33 degrees

            ii. Rock debris–40 degrees

         b. When the angle of repose of a material is exceeded, mass wasting may occur

   C. Removal of vegetation

      1. Fire

      2. Clearing of land for human use

   D. Ground vibrations produced by earthquakes may trigger mass wasting

III. Types of Mass Wasting

   A. Falls–material free falls

      1. Rockfall

         a. Material of any size falls due primarily to frost wedging

         b. Angular rock debris

            i. Accumulates at angle of repose at base of slope to form talus slopes

         c. Sudden, common in mountainous areas

B. Slides–material moves along a surface of failure

1. Slump (Fig. 8-1)

   a. Material moves along a curved surface of failure

   b. Commonly caused by undercutting of slopes by running water or waves

2. Rockslide (Fig. 8-2)

   a. Material moves along a more planar surface of failure

   b. Very sudden, common in mountainous areas

   c. May be triggered by earthquakes or melting snow

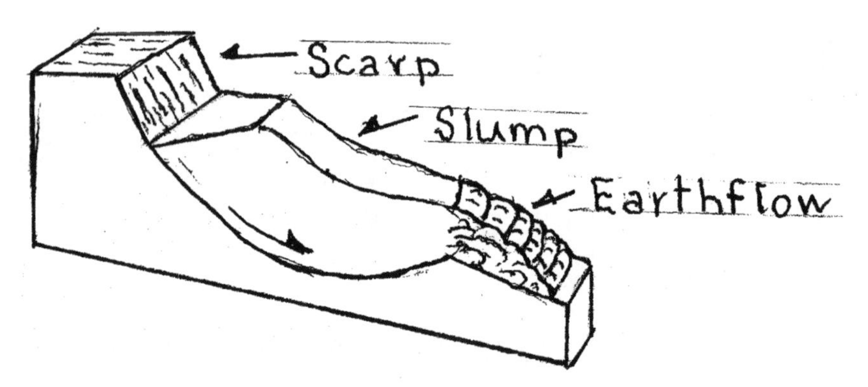


**Figure 8-1** Slump and Earthflow


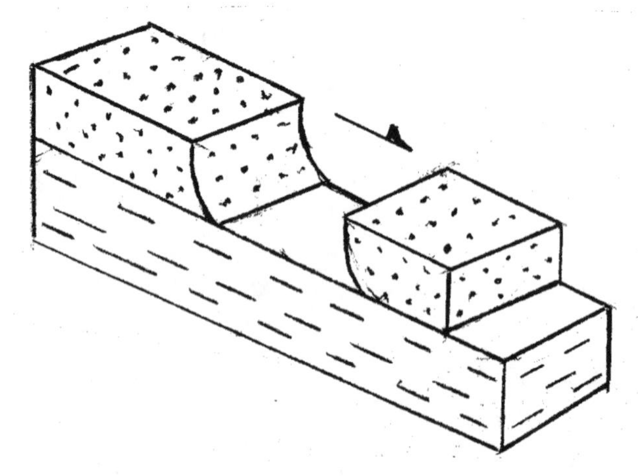

**Figure 8-2** Rockslide

C. Slides—material moves as a viscous fluid

1. Mudflow (Fig. 8-3)

   a. Water mixed with fine sediment and rock debris

2. Lahar

   a. Water mixed with fine volcanic material

3. Earthflow (Fig. 8-1)

   a. Tongue-shaped flow of fine material

   b. Occur on grassy slopes in humid (wet temperate) areas

4. Creep (Fig. 8-4)

   a. Intermittent movement of fine earth material downslope

   b. Caused by alternate wetting/drying or freezing/thawing

   c. Recognized by occurrence of bent fences, bent trees, etc.

5. Solifluction

   a. Summer thawing of active zone in area of permafrost (50–70 degrees latitude)

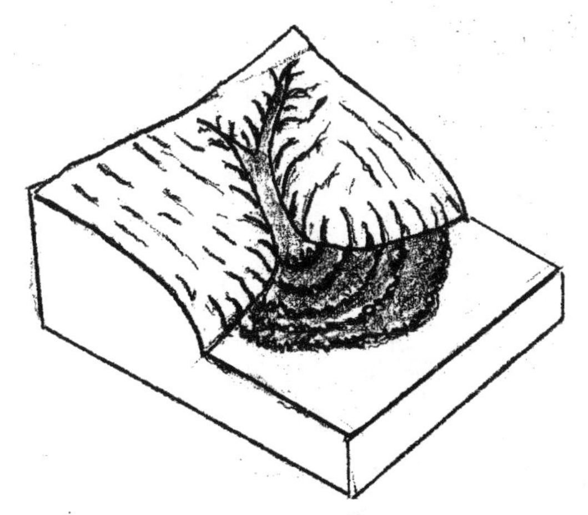

**Figure 8-3** Mudflow
*© A. Troell, 2008*

**Figure 8-4** Hillside Creep
*© A. Troell, 2008*

# NOTES

# CHAPTER 9

## Streams

I. Streams

   A. General term for rivers, creeks, etc.

   B. Streams are very important erosional agents

   C. Streams erode, transport, and deposit material

   D. Streams do most of their work during flooding

   E. Longitudinal profile of a stream (Fig. 9-1)

II. Stream Characteristics

   A. Velocity

      1. Distance water travels in given unit of time–mi/hr or km/hr

      2. Velocity determines how much material a stream can erode and transport

      3. Velocity is influenced by

         a. Gradient

            i. Slope of stream channel

            ii. Vertical drop over horizontal distance–ft/mi, m/km

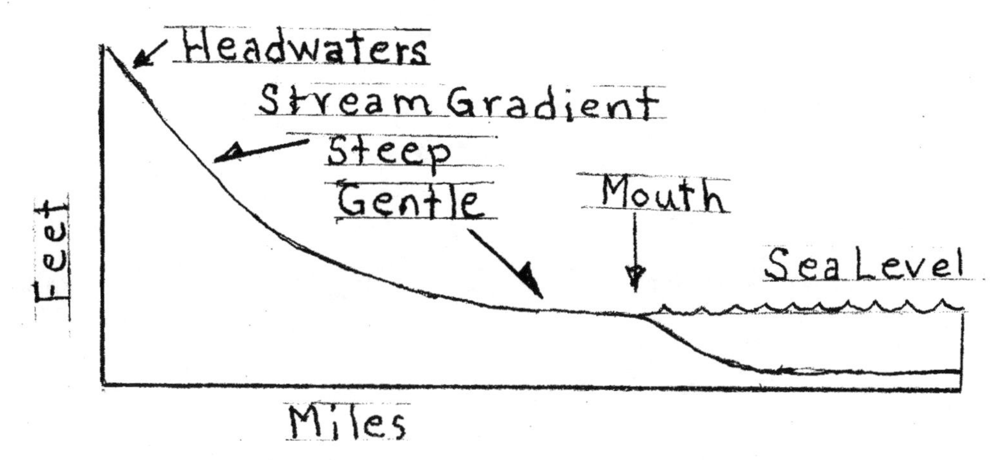


**Figure 9-1** Longitudinal Profile of a Stream
*© A. Troell, 2008*

b. Discharge

i. Volume of water flowing past certain point–$ft^3/sec$, $m^3/sec$

Channel characteristics → c. Size and shape of the channel and size of material in channel

i. Large boulders impede the velocity of the stream

III. Work of Streams

A. Erosion

1. Lifting and removal of material

2. Hydraulic action

a. Force of running water loosens and lifts material

3. Abrasion–material transported along channel has an abrasive effect

down cutting - channel getting deeper

B. Transportation

1. Dissolved load

a. Ions transported in solution

b. Derived primarily from chemical weathering

2. Suspended load

a. Sediment transported in the water column

b. Usually clay, silt, and fine sand

3. Bedload

intermitten - doesnt move constantly

a. Material transported along channel bottom

b. Sand and larger sized sediment

C. Deposition

1. Alluvium

a. Sediment deposited by streams

b. Critical settling velocity

i. Largest particles settle first

ii. Clay-sized particles won't settle until there is little turbulence to keep sediment suspended

2. Deltas

a. Depositional features formed when streams enter lake or ocean

b. Have an overall triangular shape

*Competence- maximum size particle stream can transport velocity comp.

*capacity- total amount of sediment stream can transport - related to discharge

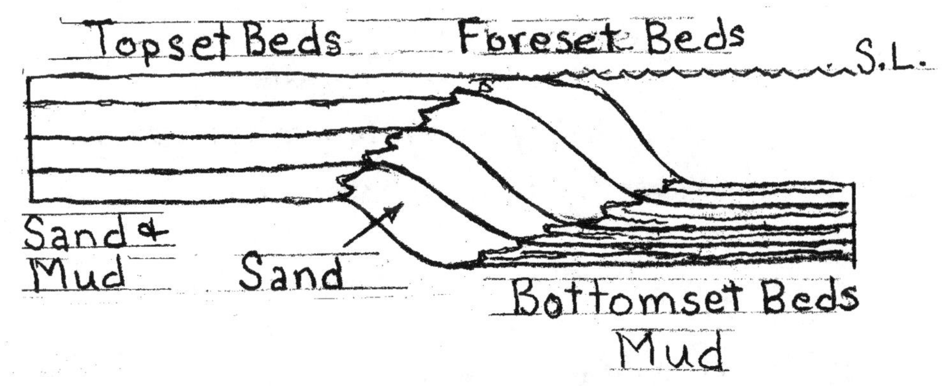


**Figure 9-2** River Delta
*© A. Troell, 2008*

c. Delta deposits (Fig. 9-2)

i. Topset beds

ii. Foreset beds

iii. Bottomset beds

Prograde- Built outward.

IV. Base Level

A. Lowest level to which a stream can erode its channel

B. Ultimate base level (Fig. 9-1)

1. Sea level

a. Streams cannot erode their channels below sea level because they need some gradient to flow

C. Temporary base level

1. May be formed by lakes or resistant layers of rock along the course of a stream

2. Limits the amount of downcutting that may occur upstream from a dam or resistant rock layer

V. Stream Valleys

A. Narrow stream valleys (youthful stage)–(Fig. 9-3)

1. Form where streams are high above base level

2. Characteristics

a. V-shaped valley

b. Downcutting is dominant

i. Stream is cutting its valley deeper

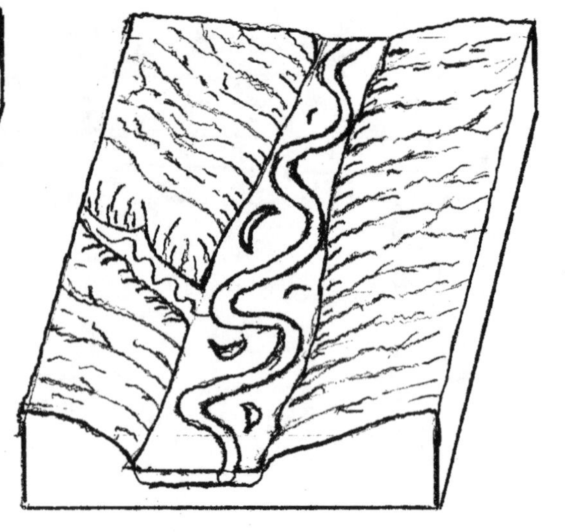

**Narrow Stream Valley**
Steep Gradient
Down Cutting
Rapids & Waterfalls
Lacks Floodplain
Youthful Stream

**Wide Stream Vally**
Gentle Gradient
Eroding Laterally
R & W Eroded
Has Floodplain
Mature Stream

**Figure 9-3** Narrow and Wide Stream Valley
*© After A. N. Strahler*

c. Rapids and waterfalls

d. Lacks a floodplain

B. Wide stream valleys (mature to old age stage)

1. Form where streams are close to base level

2. Features of meandering streams (Fig. 9-4)

a. Wide stream valley

b. Dominant work is lateral (side-to-side) erosion

c. Floodplain—flat valley floor between valley walls

d. Meanders—looping bends in streams

e. Point bars—deposits of sand on the inside of the meander (slowest velocity)

f. Cut banks—highest velocity is on the outside of the meander—erosion occurs

g. Oxbow lakes—form when a river cuts off a meander

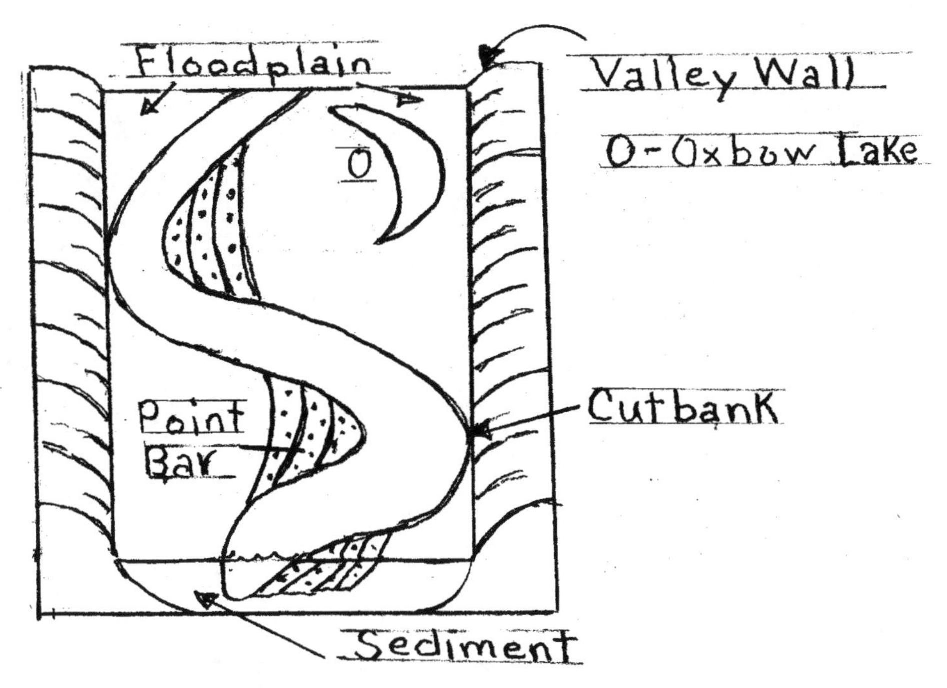


**Figure 9-4** Meandering Stream
*© A. Troell, 2008*

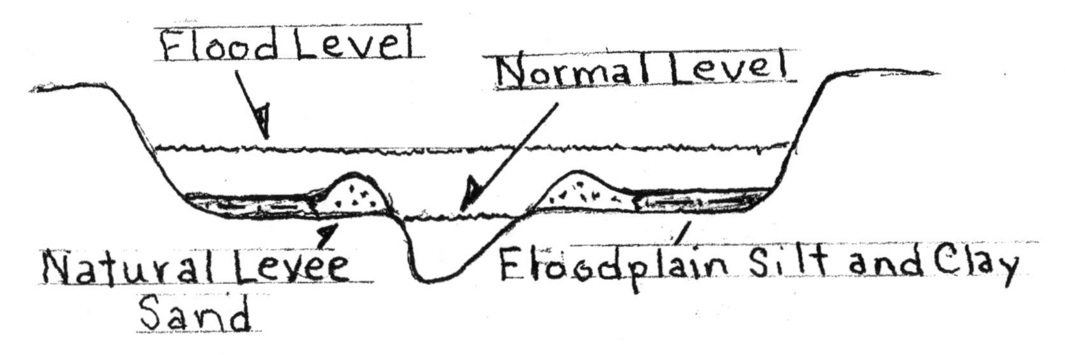


**Figure 9-5** Flood and Normal Water Levels
*© A. Troell, 2008*

h. Natural levee (Fig. 9-5)

  i. Deposit of fine sand along stream channel

  ii. Form during successive floods

i. Braided streams (Fig. 9-6)

  i. Bedload streams

  ii. Network of channels

sand gravel bars
Found in front of melting glaciers

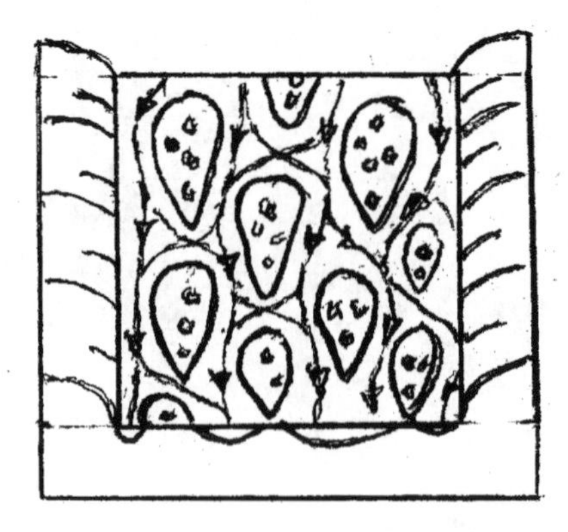


**Figure 9-6** Braided Stream Sand and Gravel Bars
*© A. Troell, 2008*

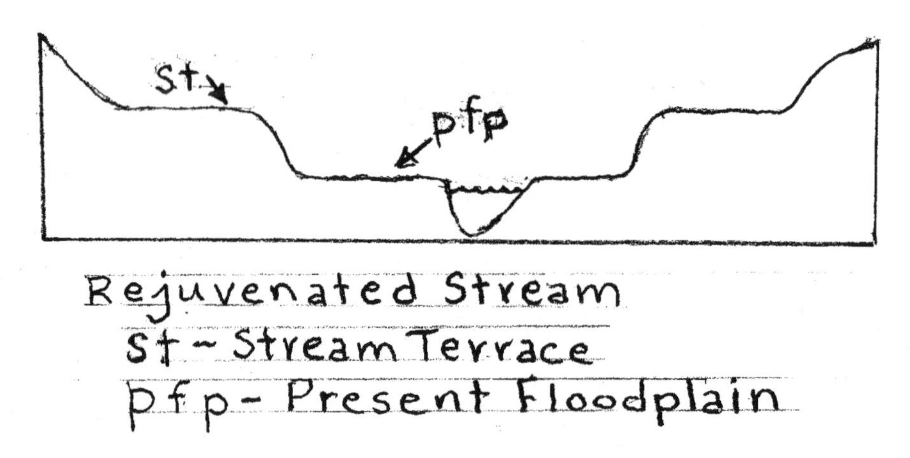


**Figure 9-7** Rejuvenated Stream, st-Stream Terrace, and pfp-Present Floodplain
*© A. Troell, 2008*

C. Rejuvenation

1. Stream terraces–Older, higher abandoned floodplain (Fig. 9-7)

2. Entrenched stream meanders–streams that meander, but have no floodplain

VI. Drainage Basins and Divides

A. Drainage basin–Area that contributes water to a stream (Fig. 9-8)

B. Divide–Separates drainage basins (Fig. 9-9)

C. Types of drainage patterns in drainage basins (Fig. 9-10)

1. Dendritic pattern

a. Tributaries join in a treelike branching pattern

b. Forms in areas underlain by fairly uniform rock types

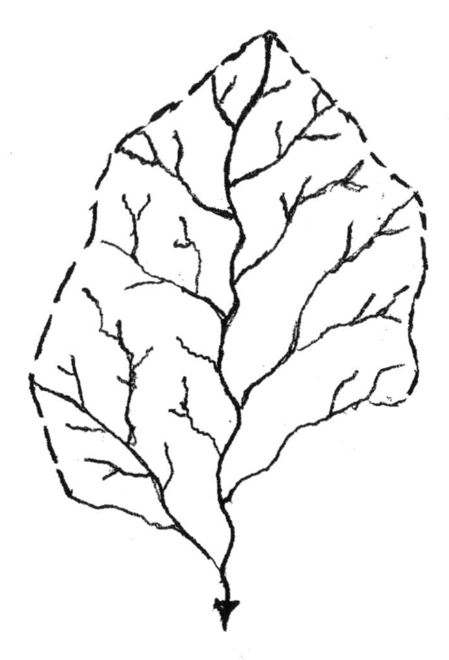

**Figure 9-8** Drainage Basin Showing Channel Network
*© A. Troell, 2008*

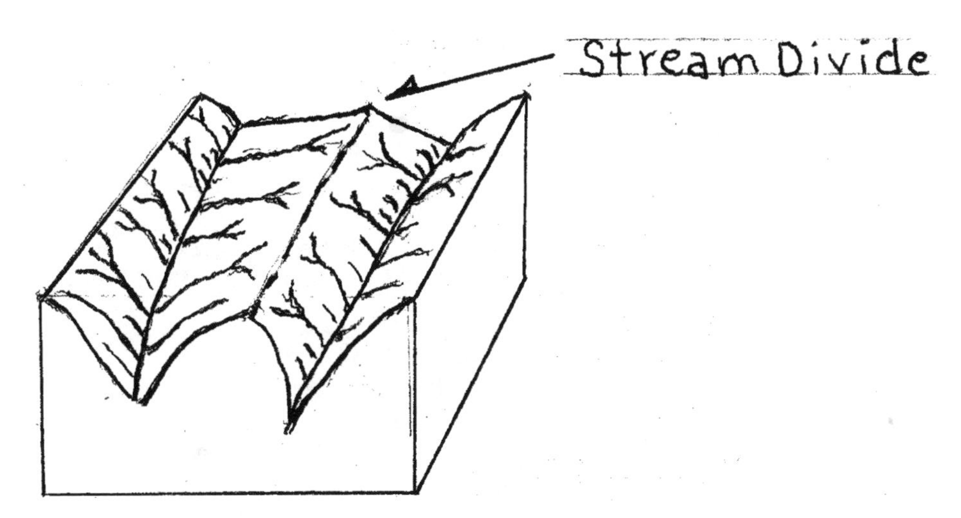


**Figure 9-9** Stream Divide
*© A. Troell, 2008*

2. Rectangular pattern
   a. Streams join at right angles
   b. Pattern is controlled by joints in the underlying bedrock
3. Radial pattern
   a. Streams diverge in all directions from a high area such as a volcano
   b. Pattern is controlled by topography

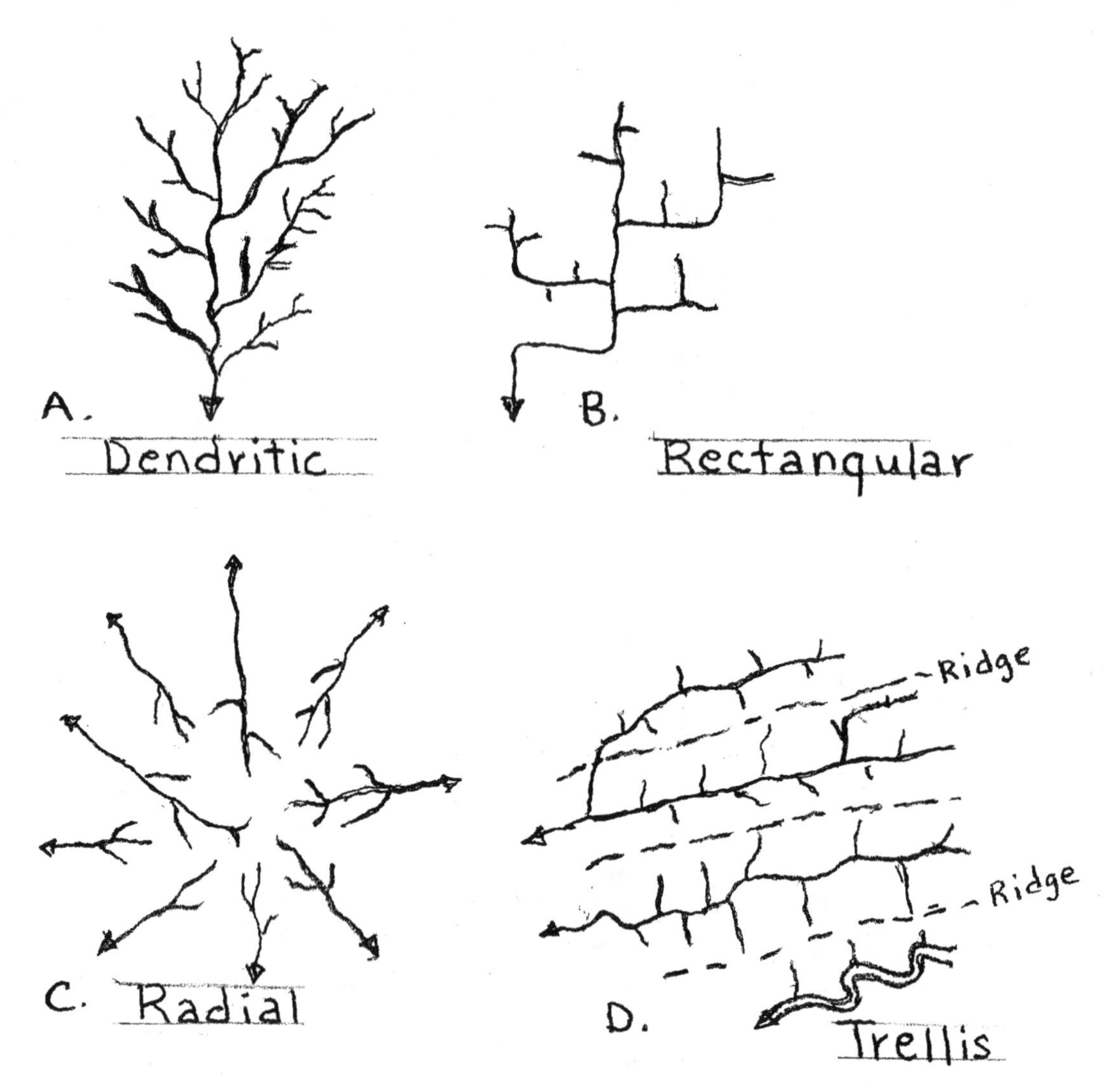


**Figure 9-10** Stream Drainage Patterns
A. Gently Inclined Strata; B. Fractured Carbonate Strata; C. Eroded Dome; D. Folded Strata
*© A. Troell, 2008*

4. Trellis pattern
    a. Tributary streams are parallel
    b. Forms where the underlying beds are folded

VI. Floods and Flood Control
  A. Flooding is caused by weather—low pressure systems or heavy spring snow melts
  B. Structures created to control flooding
    1. Artificial levees
        a. Built on top of natural levees or where no levees exist
        b. Keeps water channelized
        c. Increases flooding downstream

2. Flood-control dams
    a. Built to trap floodwaters and slowly release them
    b. Alter streams in various ways
        1. Effect ecology of streams
        2. Reservoirs cover large areas of usable land
3. Artificial cutoffs
    a. Meanders are cutoff by humans
    b. Increases gradient of stream

# NOTES

# CHAPTER 10

## Groundwater

I. Groundwater–Water that occurs in soil and rock

II. Groundwater Basics

   A. Porosity–storage capacity of rocks or sediment

   B. Permeability–how fast water moves through rocks or sediment

   C. Aquitards ("water-closed")–impermeable rocks that prevent groundwater movement

   D. Aquifers ("water-bearing")

      1. Permeable rocks that store groundwater

      2. Unconfined aquifers (Fig. 10-1)

         a. Zone of aeration–layer in which pore spaces are filled with air

         b. Zone of saturation–layer in which all the pore spaces are filled with water

         c. Water table

            i. Top of the zone of saturation

            ii. Fluctuates during the year

            iii. Mimics overlying topography (ground surface)

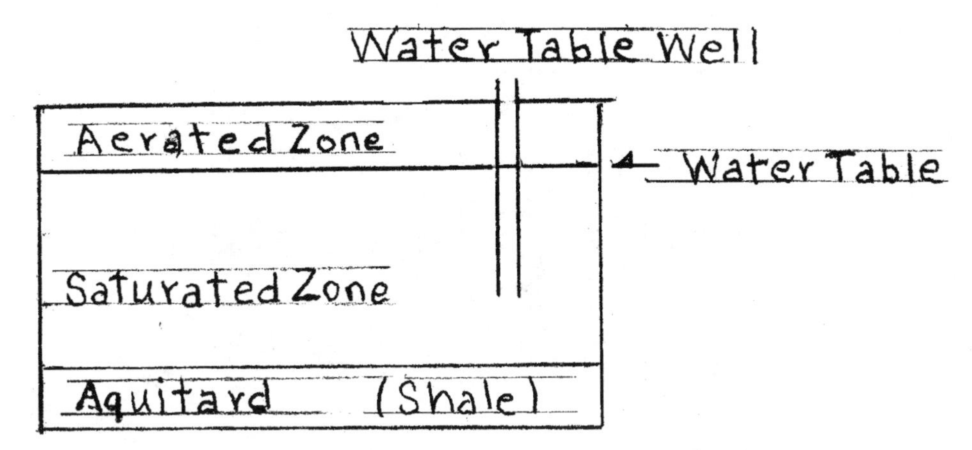


**Figure 10-1** Unconfined Aquifer
*© A. Troell, 2008*

d. Unconfined aquifers have an aquitard below, but not above

e. Water-table wells are drilled into unconfined aquifers

i. When water is withdrawn from an unconfined aquifer, a cone of depression forms around the well

f. The Edwards Aquifer is unconfined in the recharge zone

g. Perched water table

i. Localized water table above the main water table

ii. Form where there is an aquitard in the zone of aeration that traps water above the main water table

3. Confined aquifers (Fig. 10-2)

a. Have aquitards both above and below the aquifer

b. Artesian-pressure surface

c. Artesian wells–water in the aquifer is under pressure and will rise to the level of the artesian-pressure surface

d. The Edwards Aquifer is confined underneath San Antonio College

III. Springs and Geysers

A. Springs–form where the water table intersects the ground surface

B. Hot springs–form where water is warmer than mean annual air temperature

C. Geysers–intermittent hot springs in which hot water erupts followed by steam

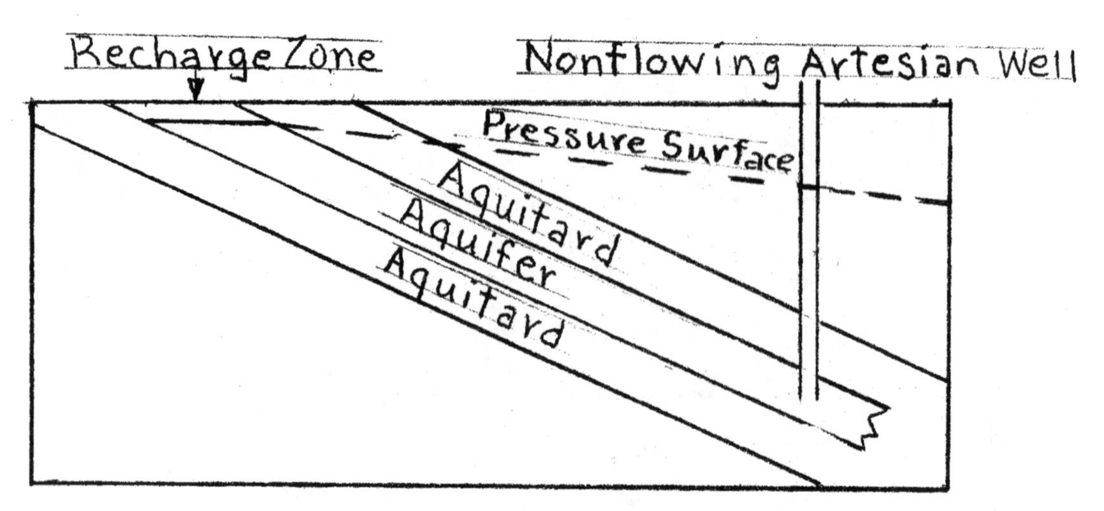


**Figure 10-2** Confined Aquifer

*© A. Troell, 2008*

IV. Environmental Problems

A. Mining an aquifer

1. Withdrawing groundwater faster than recharge occurs

2. High Plains aquifer–North Texas to South Dakota

B. Ground subsidence

1. Water helps hold grains apart

2. Withdrawal of water allows grains to compact more closely

3. San Joaquin Valley

a. Withdrawal of groundwater between 1925 and 1975 caused the ground surface to subside up to 28 feet in places

C. Groundwater pollution–sewage, chemical discharge, landfills, etc.

V. Caves and Karst Topography (Fig. 10-3)

A. Most caves are formed at or slightly below the water table in the zone of saturation in limestones

B. Deposition of the travertine (limestone) occurs in the zone of aeration

C. Karst topography = House fell in hole.

1. Surface features that form in humid areas underlain by limestone

a. Sinkholes–form where the roof of a cave collapses (slowly or quickly), leaving an opening at the ground surface

b. Disappearing streams–form where a stream flows into a sinkhole

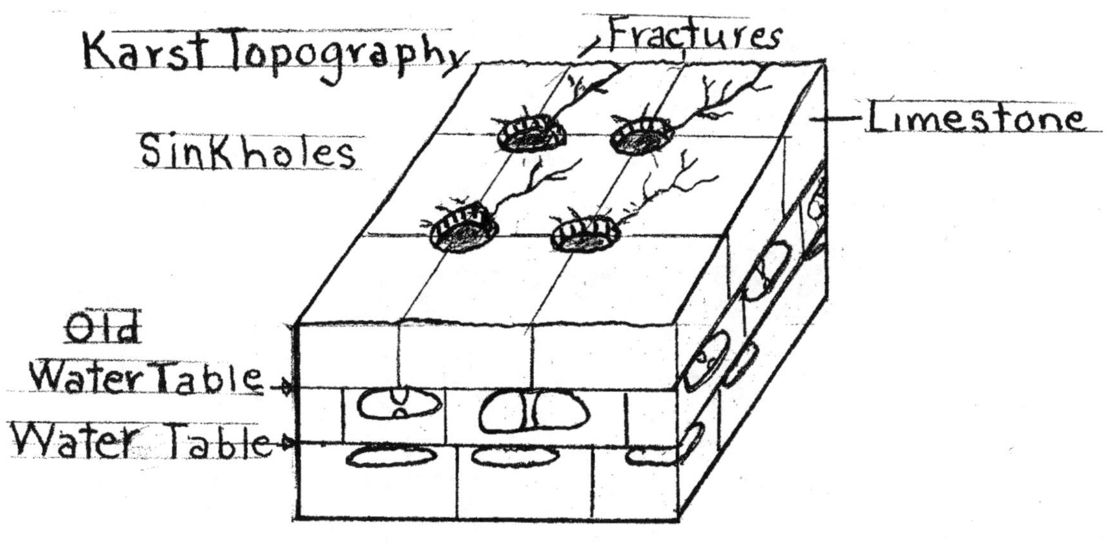


**Figure 10-3** Formation of Caves and Karst
*© A. Troell, 2008*

# NOTES

# CHAPTER 11

## Glaciers

I. Glacier

 A. Large mass of ice that accumulates over 100s–1000s of years

 B. Form from the accumulation of snow

  1. Snow is compacted and recrystallized to form glacial ice

 C. Most glaciers move very slowly

 *D. Glaciers can erode, transport, and deposit large amounts of material

 E. Glacial action is responsible for shaping many types of landforms

 *F. Glaciers cover approximately 10% of Earth's surface today

II. Types of Glaciers

 A. Valley glaciers (Fig. 11-1)

  1. Form in mountainous areas and move down stream valleys

  2. Found today on all continents except Australia

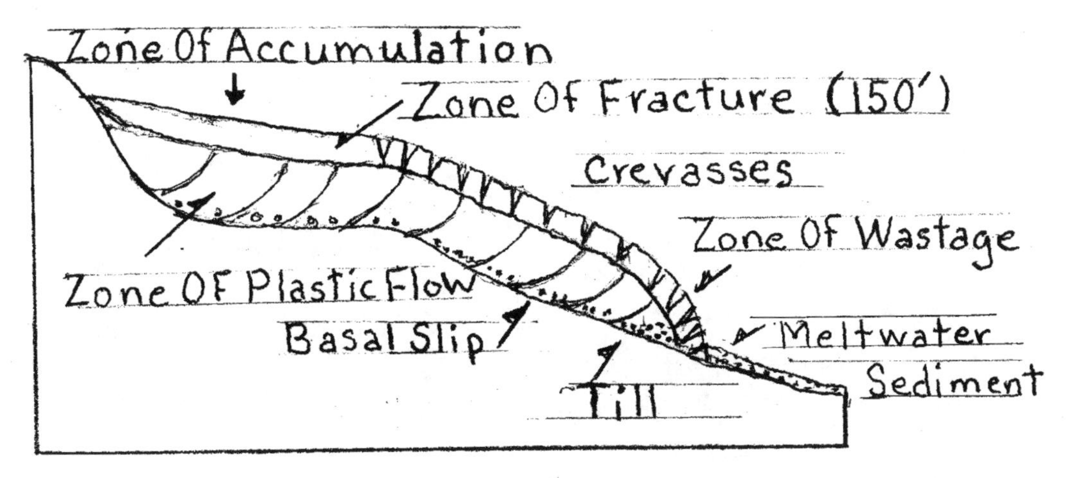


**Figure 11-1** Valley Glacier–Cross Section
*© A. Troell, 2008*

B. Ice sheets (continental glaciers)—Fig. 11-2

1. Glaciers that move outward in all directions from one or more centers of accumulation

2. Today found only in Antarctica (maximum thickness ~14,000 feet) and Greenland

III. Glacial Movement (Fig. 11-1)

A. Zone of fracture—Upper ~150 feet of the glacier

B. Zone of flow

1. Glacier movement occurs here

2. Basal slip—glacier moves along underlying bedrock due to lubrication by meltwater

3. Plastic flow—ice actually flows in this zone

IV. Glacial Budget (Figs. 11-1 and 11-2)

A. Zone of accumulation

1. Snow accumulation and ice formation occur here

2. Ice is added to the glacier

B. Zone of wastage

1. All the snow from the previous winter and some of the glacial ice are lost

a. Melting

b. Calving—pieces of the glacier break off when a glacier flows into a lake or the ocean

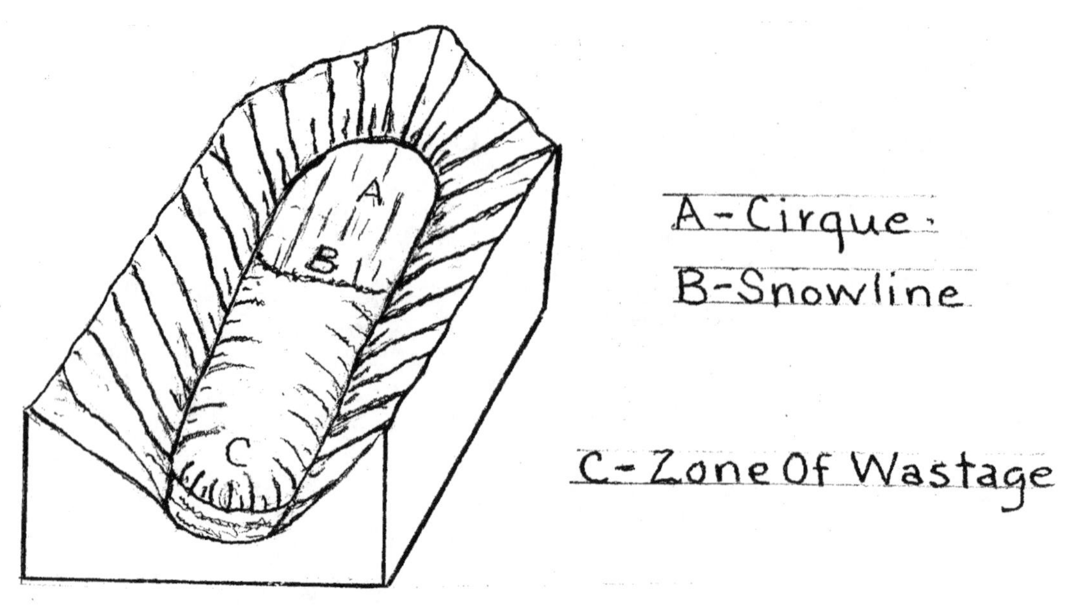


**Figure 11-2** Valley Glacier—Aerial View
*© A. Troell, 2008*

C. Glacial budget

1. Terminus–end of valley glacier or edge of an ice sheet
2. When accumulation > wastage, the terminus advances
3. When wastage > accumulation, the terminus retreats
4. When accumulation = wastage, the terminus remains stationary

V. Glacial Erosion

A. Plucking–pieces of bedrock freeze to the bottom of the glacier and are pulled out as the glacier moves

B. Abrasion

1. Abrasive effect of particles transported along the bottom and sides of the glacier
2. Glacial striations–scratches made on underlying bedrock as large particles are moved along bottom of glacier
3. Rock flour
    i. Pulverized rock material (silt and clay-sized) created as rocks grind against each other during transport
    ii. May be picked up by wind and deposited as loess

VI. Erosional Features Created by Valley Glaciers

A. U-shaped glacial trough–valley glaciers widen, deepen, and straighten stream valleys

B. Cirques–bowl-shaped depressions at the head of glacial valley

C. Hanging valley

1. Smaller tributary glaciers can't erode their valleys as deeply as the main glacier
2. When the glaciers melt, the smaller valleys are left above the main valley
3. Waterfalls form where water flows from hanging valleys

D. Aretes–knifelike ridges that separate glacial valleys or cirques

E. Horns–pyramidlike peaks that form in areas where three or more cirques surround a peak

F. Fjords–drowned glacial valleys along coasts

VII. Glacial Deposition

A. Drift–general term for sediment deposited by glaciers

1. Till–poorly sorted glacial material deposited when glacial ice melts
2. Stratified drift–glacial deposits transported and deposited by meltwater

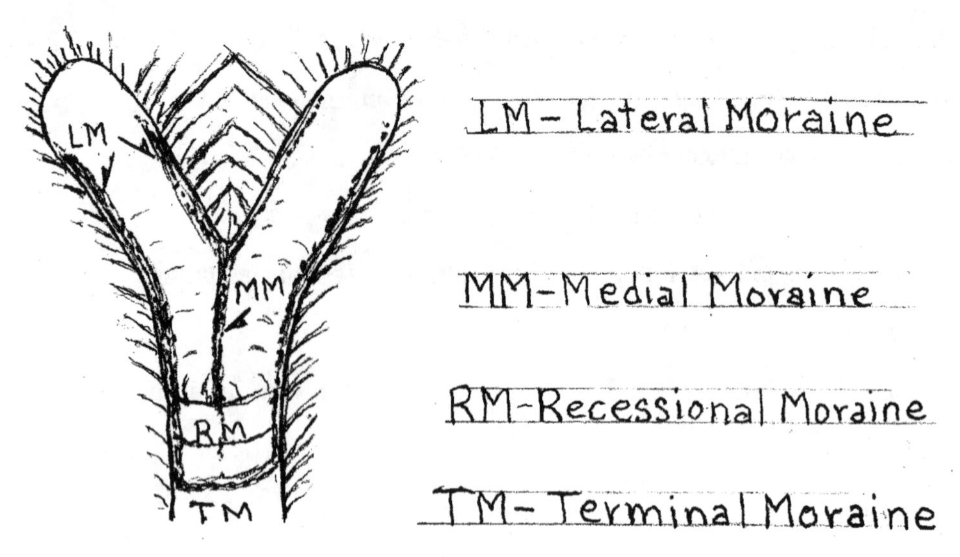


**Figure 11-3** Till Deposits
*© A. Troell, 2008*

B. Moraines (Fig. 11-3)

1. Layers or ridges of till
2. Lateral moraine–material transported along sides of a valley by a valley glacier; forms a lateral moraine when deposited
3. Medial moraine–formed where two lateral moraines join in a valley glacier
4. End moraines
    a. Form at the terminus of a valley glacier or ice sheet
    b. Terminal end moraine–form when the terminus of a glacier is stationary (Fig. 11-5)
    c. Ground moraine–layer of till deposited as a glacier retreats (Fig. 11-5)
    d. Recessional moraine–form when the terminus of a retreating glacier becomes stationary (Fig. 11-3)

C. Valley trains and outwash plains

1. Outwash plains–stratified drift deposited in front of an ice sheet (Fig. 11-5)
2. Valley trains–stratified drift deposited beyond terminus of valley glaciers

VIII. Ice Sheets

A. Often produce rolling topography

B. Features associated with retreating ice sheets (continental glaciers)–(Figs. 11-4 and 11-5)

1. Drumlins (Fig. 11-5)

   a. Asymmetrical hills composed of till

   b. Probably molded in the zone of flow

2. Eskers–sinuous ridges of stratified drift deposited by meltwater flowing under the edge of glaciers (Fig. 11-5)

3. Kettles–depressions formed when a block of ice is left behind by a retreating glacier and buried in drift (Figs. 11-4 and 11-5)

4. Braided streams (Fig. 11-4)

   a. Meltwater flowing from glaciers carries large load of sediment

   b. Gravel and sand bars are deposited in the channels

   c. When water level is low, water has to flow around the bars producing a network of channels

5. Terminal end and recessional moraines

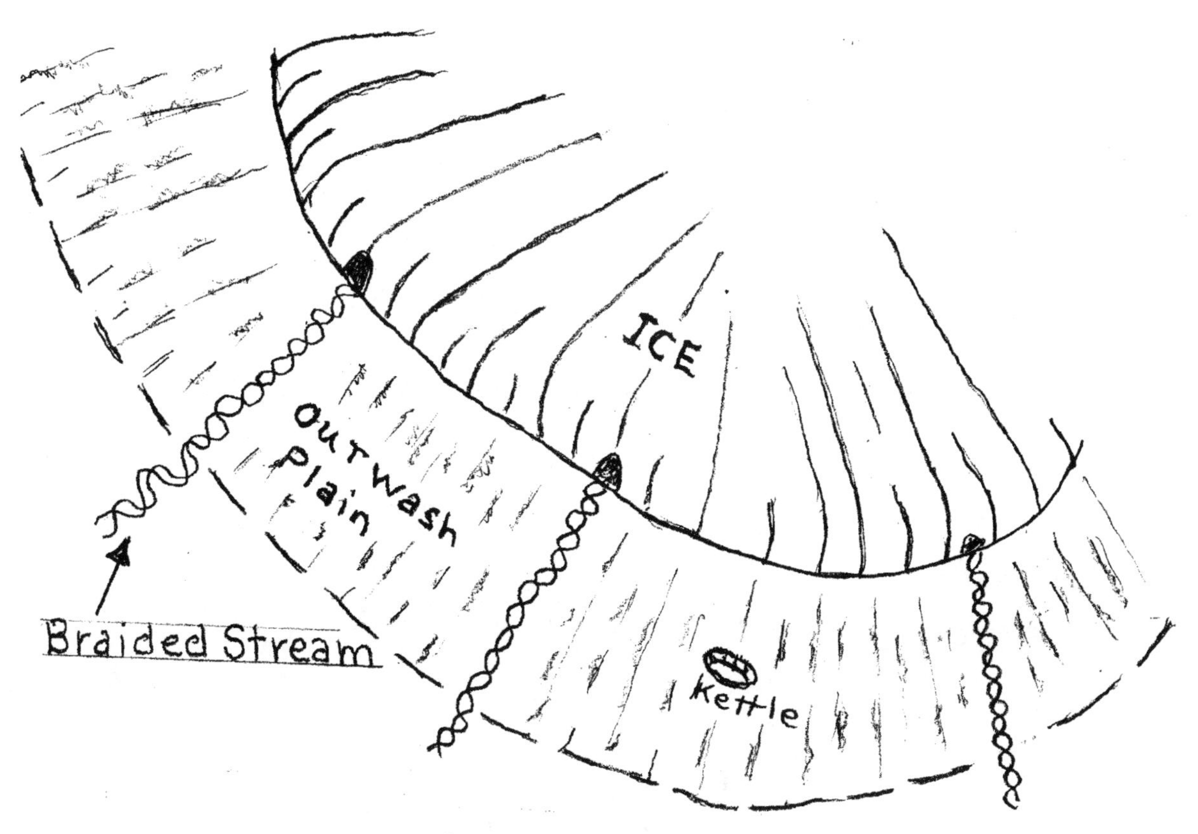


**Figure 11-4** Continental Glacier

*© A. Troell, 2008*

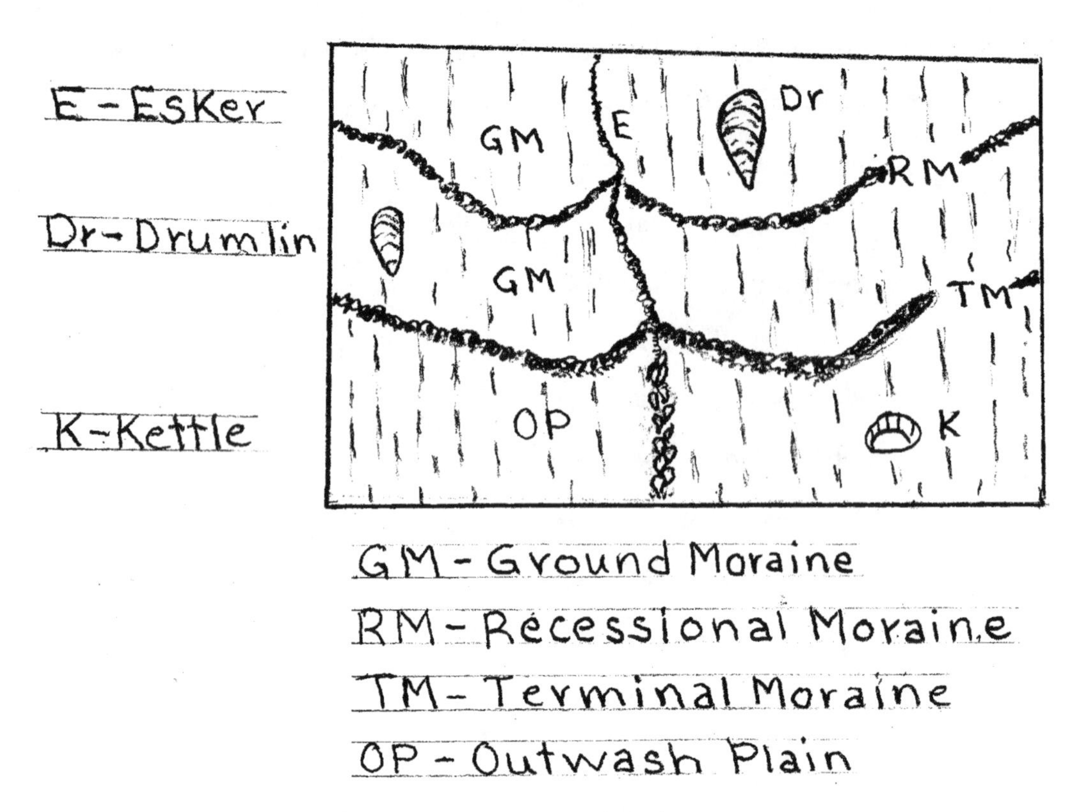


**Figure 11-5** Retreating Glacier
*© A. Troell, 2008*

IX. Pleistocene Epoch

A. Pleistocene Epoch

1. 1.6 my–10,000 years
2. Ice sheets covered approximately 30% of Earth's surface
3. Evidence for four glacial advances and retreats on land
4. Evidence from ocean floor sediment indicates at least 20 warming/cooling cycles occurred

B. Effects of glaciation

1. Sea level changes
   a. Occur when glaciers advance and retreat
   b. Sea level was about 450 feet lower than today at times during Pleistocene
2. Isostatic rebound
   a. Crust was depressed by weight of ice sheets
   b. When ice sheets retreated, crust slowly rebounded
      i. Some areas are still rebounding from Pleistocene glaciation

3. Some mammals grew to unusually large sizes
    a. Mammoths, ground sloths, Irish elk, glyptodonts, etc.
4. Pluvial lakes
    a. Climates during the Pleistocene were cooler and wetter
    b. Permanent lakes formed in areas such as Death Valley
    c. Great Salt Lake is a remnant of much larger lake that formed during Pleistocene

# NOTES

# CHAPTER 12

## Deserts

I. Introduction

  A. Deserts (arid)–receive less than 10 inches of rain per year (average)

  B. Steppes (semiarid)

    1. Receive 10–25 inches of rain per year (average)

    2. Generally separate deserts from more humid environments

  C. Desertification–in some areas, deserts are expanding due to human actions

II. Types of Deserts

  A. Polar deserts

    1. Intensely cold, perpetual snow cover; no liquid water

    2. Antarctica, interior of Greenland

  B. Subtropical deserts

    1. Form at 30 degrees North or South of equator in zones of subsiding air

    2. Large daily temperature variation

    3. Sahara desert, Great Australian desert

  C. Mid-Latitude deserts

    1. Rain shadow deserts

      a. Form where mountain ranges create a barrier to moist air

        i. Humid air rises and rainfall occurs on windward side, air warms when it descends on the opposite (leeward) side of mountains, and clouds don't form

      b. Great Basin, North America

    2. Interior deserts

      a. Form in the interiors of continents far from moisture of oceans

      b. Hot summers, cold winters

      c. Gobi desert in central Asia

III. Characteristics of Deserts

A. Running water is most important erosional agent in deserts

B. Mechanical weathering dominant

C. Thin, poorly developed soils

D. Ephemeral streams–streams that flow only in response to rainfall

E. Flash floods are common–a desert may receive most or all of its yearly rainfall at one time

F. Sparse, but well-adapted vegetation

IV. Landforms Developed in Dry Mountainous Areas (Figs. 12-1 and 12-2)

A. Alluvial fan–cone of debris deposited at mouth of a canyon where streams flow out onto the flat desert floor

B. Bajada–broad apron of sediment formed by coalescing alluvial fans

C. Playa lake–temporary lake in a desert

D. Playa–dry, flat lake bed that may be encrusted with salts

E. Inselberg–isolated, steep-sided erosional remnants of mountains

F. Basin and range

1. Extends from southern Oregon through Arizona into Mexico

2. Consists of about two hundred small mountain ranges separated by valleys

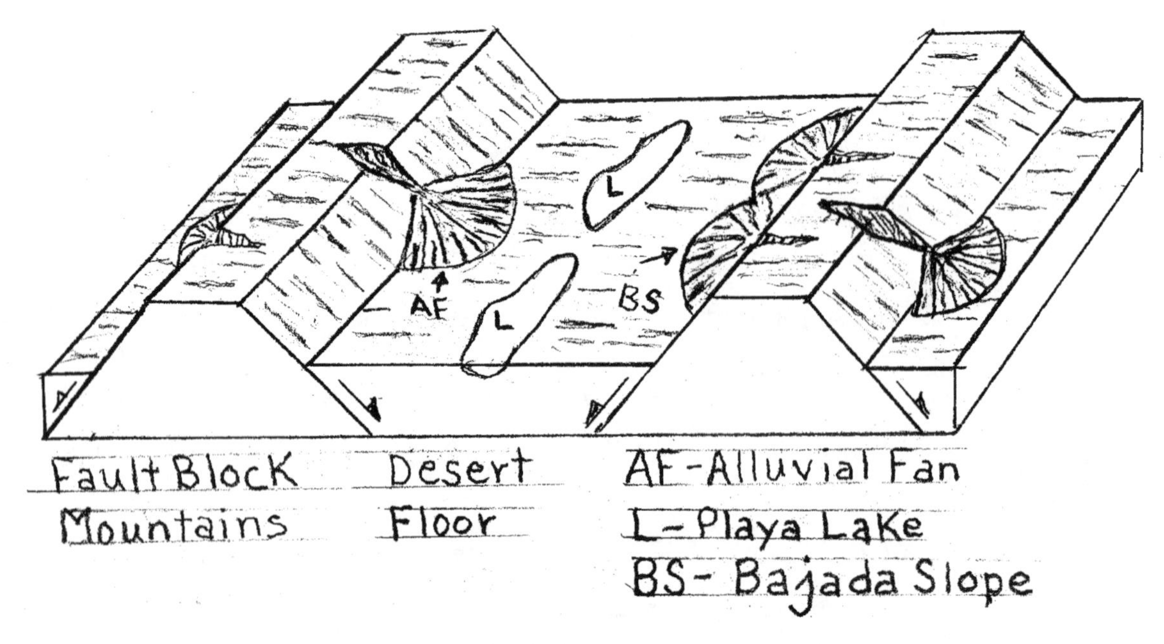


**Figure 12-1** Fault Block Mountains, Desert Floor, etc.

3. In southern Oregon and northern Nevada, the mountains are just beginning to erode
    a. Alluvial fans and playa lakes common (Fig. 12-1)
4. In southern Nevada and northern Arizona, mountains are in a later stage of erosion
    a. Valleys are beginning to fill in with material deposited from eroding mountains
        i. Bajadas common
5. In southern Arizona, the mountains are in a late stage of erosion (Fig. 12-2)
    a. Inselbergs are all that remain of the eroded mountains

V. Wind Erosion
  A. Deflation
    1. Lifting and removal of loose material
    2. Deflation lowers the surface of the land
    3. Blowouts–shallow depressions created by deflation
    4. Desert pavement
        a. Layer of coarse particles left behind as deflation removes the finer sand and silt
        b. As long as desert pavement is undisturbed, it prevents further deflation
  B. Abrasion
    1. Windblown sand cuts and polishes surfaces
    2. Ventifacts–rocks with flat sides that have been planed by windblown sand

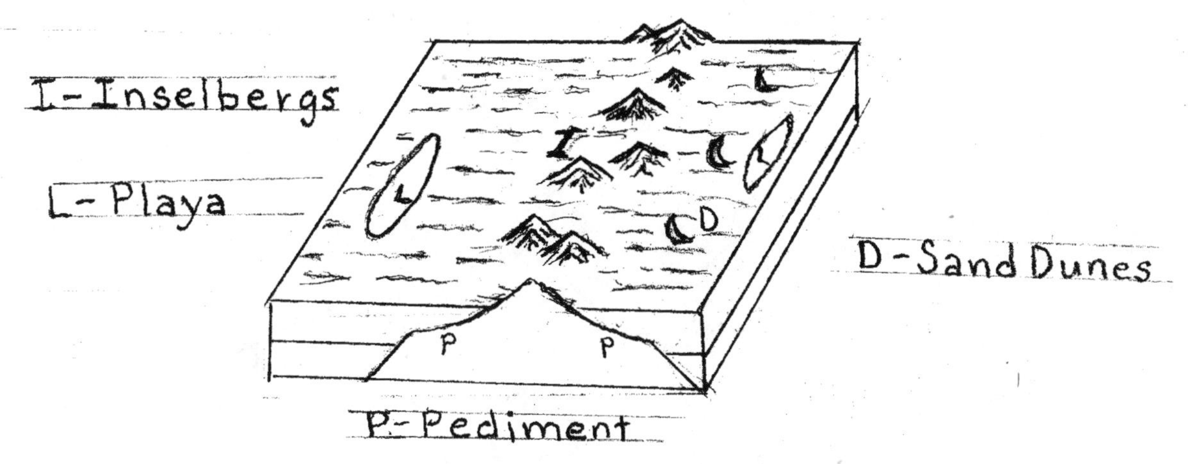


**Figure 12-2** I-Inselbergs, L-Playa, P-Pediment, D-Sand Dunes
*© A. Troell, 2008*

VI. Wind Deposits

A. Loess

1. Windblown silt and clay; tan to yellow color
2. Sources
    a. Deserts
    b. Glaciers–stratified drift

B. Sand dunes

1. Mounds or ridges of sand
2. Commonly have an asymmetrical profile
    a. Sand moves by saltation (bouncing and skipping) up the windward side
    b. Cross beds–inclined layers of sand that form on the leeward side
3. Types of sand dunes
    a. Barchan dunes (Fig. 12-3)
    b. Longitudinal dunes
    c. Transverse dunes
    d. Parabolic dunes
    e. Star dunes

Landforms Of Mountainous Deserts
Of The Southwestern United States

Figure 1 Initial Stage
Figure 2 Advanced Stage

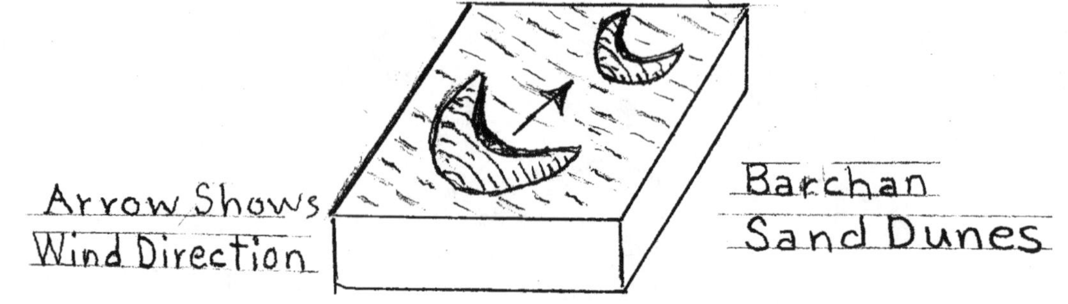


**Figure 12-3** Barchan Sand Dunes

# NOTES

# NOTES

# CHAPTER 13

## Rock Deformation

I. Rock Deformation

   A. When rocks are stressed beyond their elastic limit, permanent deformation results

   B. Types of stress (Fig. 13-1)

      1. Tensional stress–divergent plate boundaries

      2. Compressional stress–convergent plate boundaries

      3. Shear stress–transform boundaries

II. Attitude of Beds (Fig. 13-2)

   A. Dip–angle of inclination of beds measured from the horizontal

   B. Strike–line formed by the intersection of a dipping bed and the earth's surface

   C. Dip direction–direction into which a bed descends (slopes)

III. Folds

   A. Compressional stress causes shortening and thickening of the crust

   B. Most folds form during mountain building events

   C. Types of folds

      1. Anticline

         a. Upward fold in rocks

         b. Limbs dip (are inclined) away from the center of the fold

         c. Oldest rocks exposed in the center

      2. Syncline

         a. Downward fold in rocks

         b. Limbs dip toward the center of the fold

         c. Youngest rocks exposed in the center

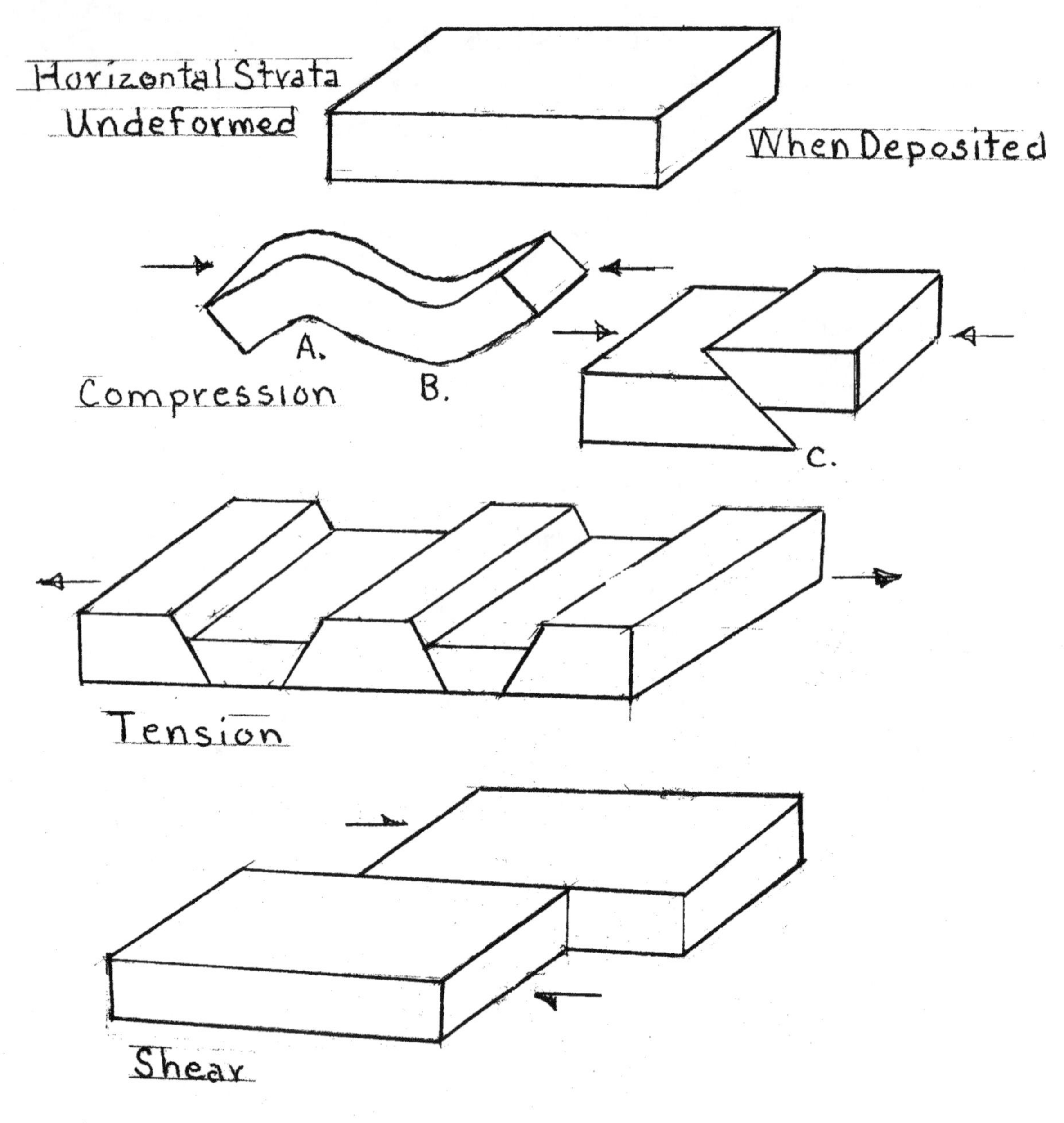


Deformation Of Strata

Compression

A. Anticline - Upfold

B. Syncline-Downfold

C. Reverse Fault

Tension

Normal Faults

Shear

StriKe-Slip Fault

**Figure 13-1** Deformation of Strata

*© After Wicander and Monroe*

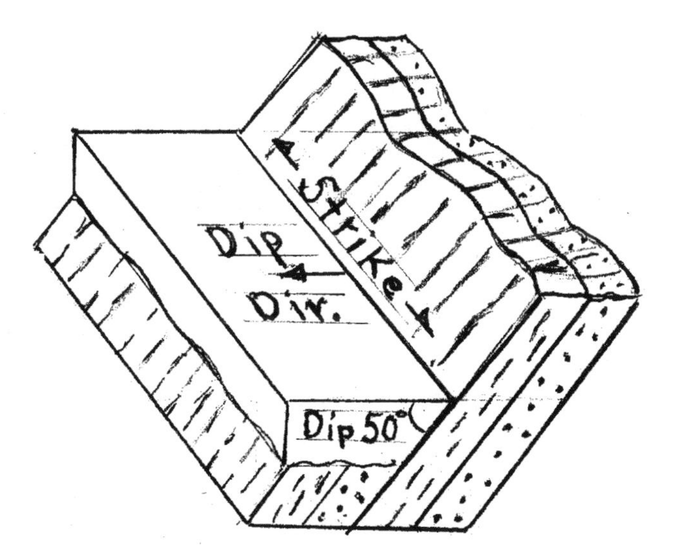


**Figure 13-2** Stike and Dip of Strata
*© A. Troell, 2008*

D. Symmetry of Nonplunging Folds (Fig. 13-3)

1. Fold axis is horizontal
2. Symmetrical anticline or syncline
    a. Limbs dip at about the same angle
3. Asymmetrical anticline or syncline
    a. One limb is inclined more steeply than the other limb
4. Overturned anticline or syncline
    a. Limbs are inclined in the same direction
5. Recumbent anticline or syncline–axial plane is horizontal
6. Erosion of nonplunging folds–repetition of beds (Fig. 13-4)

E. Plunging folds

1. Axis is inclined
2. Eroded plunging anticlines (Fig. 13-5)
    a. Oldest rocks exposed in center
    b. Rocks dip away from center
    c. "V" points in direction of plunge
3. Eroded plunging synclines (Fig. 13-5)
    a. Youngest rocks exposed in center
    b. Rocks dip toward from center
    c. "V" points away from the direction of plunge

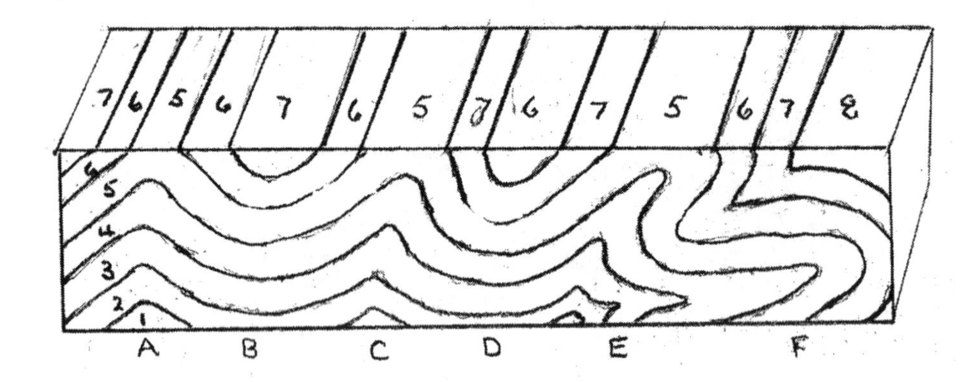


A-Symmetrical Anticline
B-Symmetrical Syncline
C-Asymmetrical Anticline
D-Asymmetrical Syncline
E-Overturned Anticline
F-Recumbent Anticline

**Figure 13-3** Symmetry of Folds

*© A. Troell, 2008*

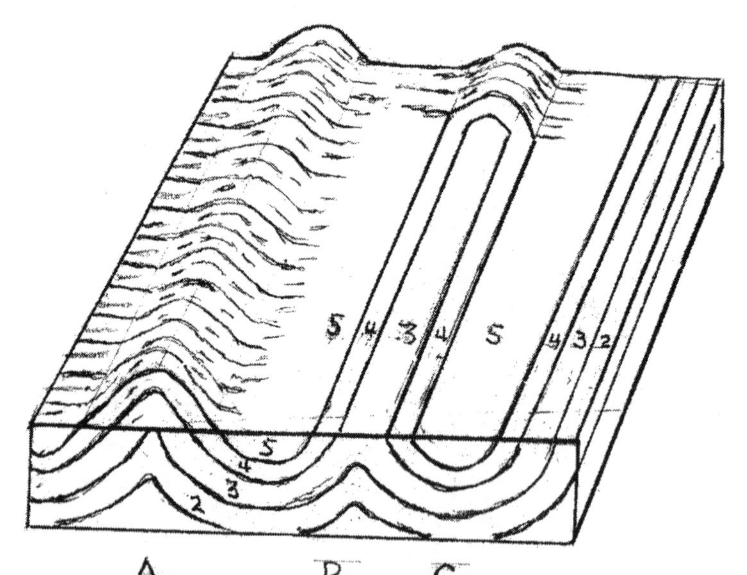


A. B. C.

Nonplunging Folds

A.-Uneroded Anticline

B.-Eroded Anticline

Older Strata In Center

C.-Eroded Syncline

Younger Strata In Center

**Figure 13-4** Stike and Dip of Strata

*© A. Troell, 2008*

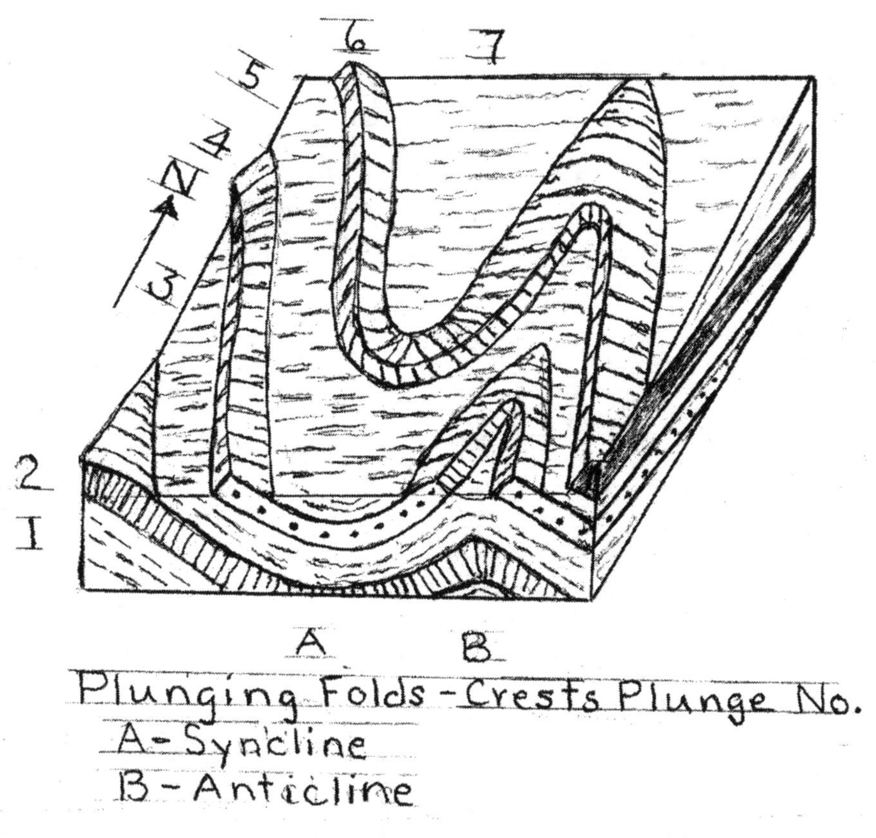


**Figure 13-5** Erosion of Plunging Folds
© *After A. N. Strahler*

F. Domes (Fig. 13-6)

1. Circular or oval equivalents of anticlines
2. Oldest beds exposed in center
3. Black Hills, South Dakota

G. Basins (Fig. 13-6)

1. Circular or oval equivalents of synclines
2. Youngest beds exposed in center
3. Michigan Basin
4. Both domes and basins normally form inside continents (not at plate boundaries)

IV. Faults

A. Fractures in the crust along which displacement has occurred

B. Dip-Slip Faults

1. Normal faults (Fig. 13-7A)
    a. Hanging wall block moves down relative to the footwall block

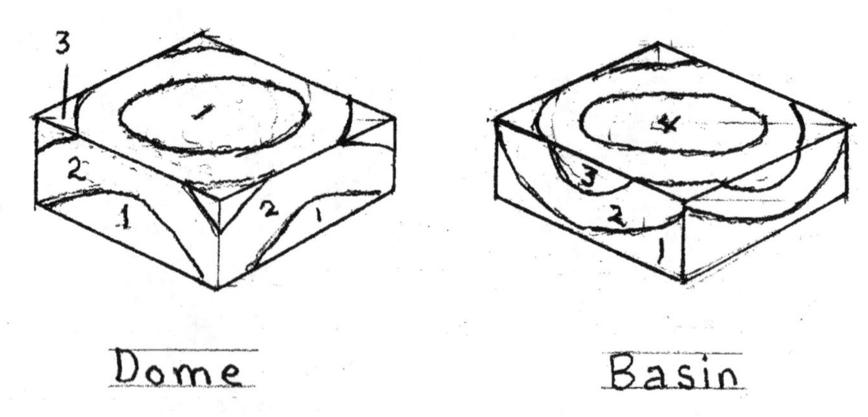


**Figure 13-6** Dome and Basin
*© A. Troell, 2008*

b. Result from tensional stress

c. Grabens–central blocks that dropped down along normal faults (Fig. 13-7C)

d. Horsts–uplifted blocks that bound grabens (Fig. 13-7C)

2. Reverse faults (Fig. 13-7B)

a. Hanging wall block moves up relative to the footwall block

b. Result from compressional stress

3. Thrust faults–low-angle (< 45 degrees) reverse faults (Fig. 13-7D)

B. Strike-slip faults (Fig. 13-7)

1. Displacement is horizontal

a. Right-lateral strike slip faults (Fig. 13-7E)

b. Left-lateral strike-slip faults (Fig. 13-7F)

2. Transform fault–type of strike-slip fault found at a plate boundary

V. Joints

A. Fractures in rocks along which no appreciable displacement has occurred

B. Sheet joints

1. Form in igneous rocks as erosion removes the overburden

2. Exfoliation domes form as a result of rocks breaking off along sheet joints

C. Columnar joints–elongate, pillarlike columns that form as basalt cools and shrinks

D. Tectonic joints–forces associated with crustal movements cause rocks to fracture (Fig. 13-7G)

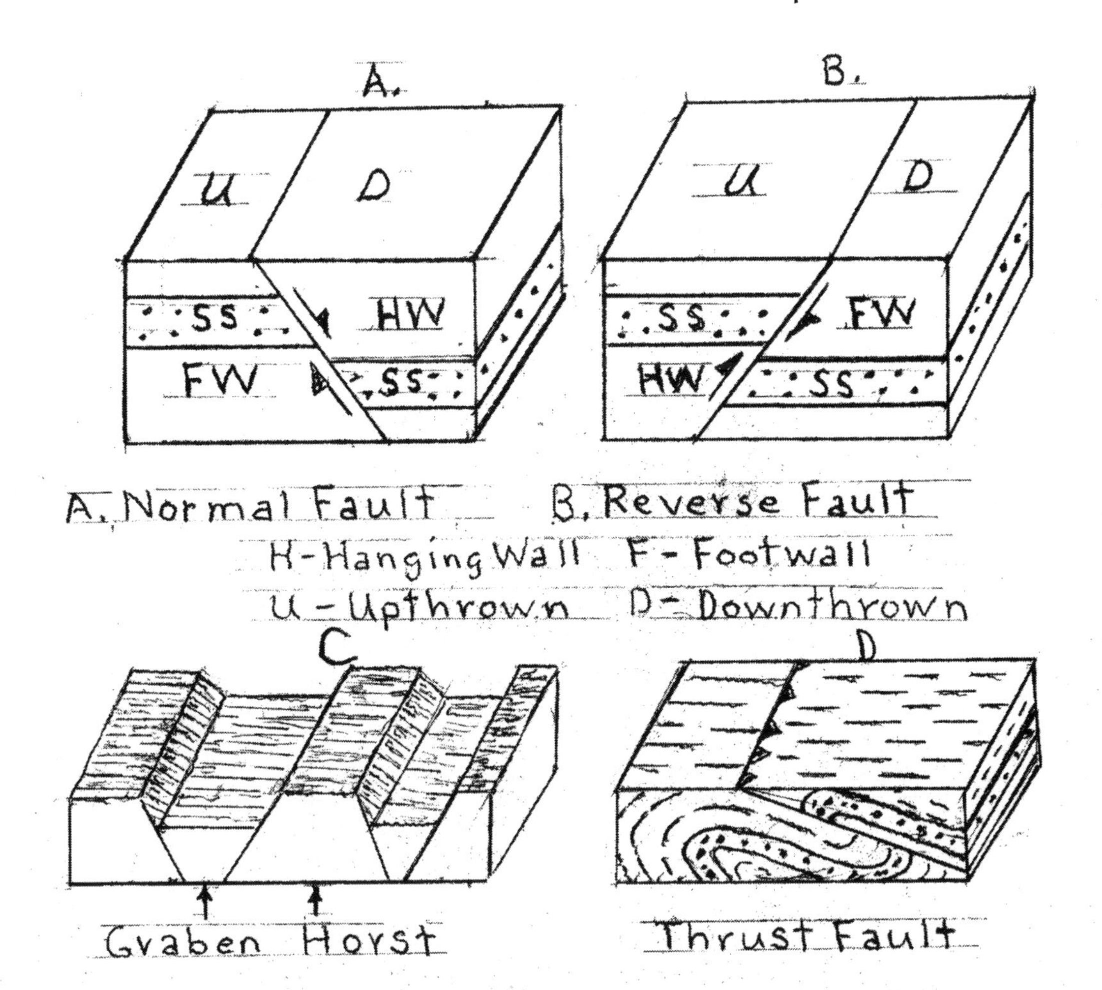


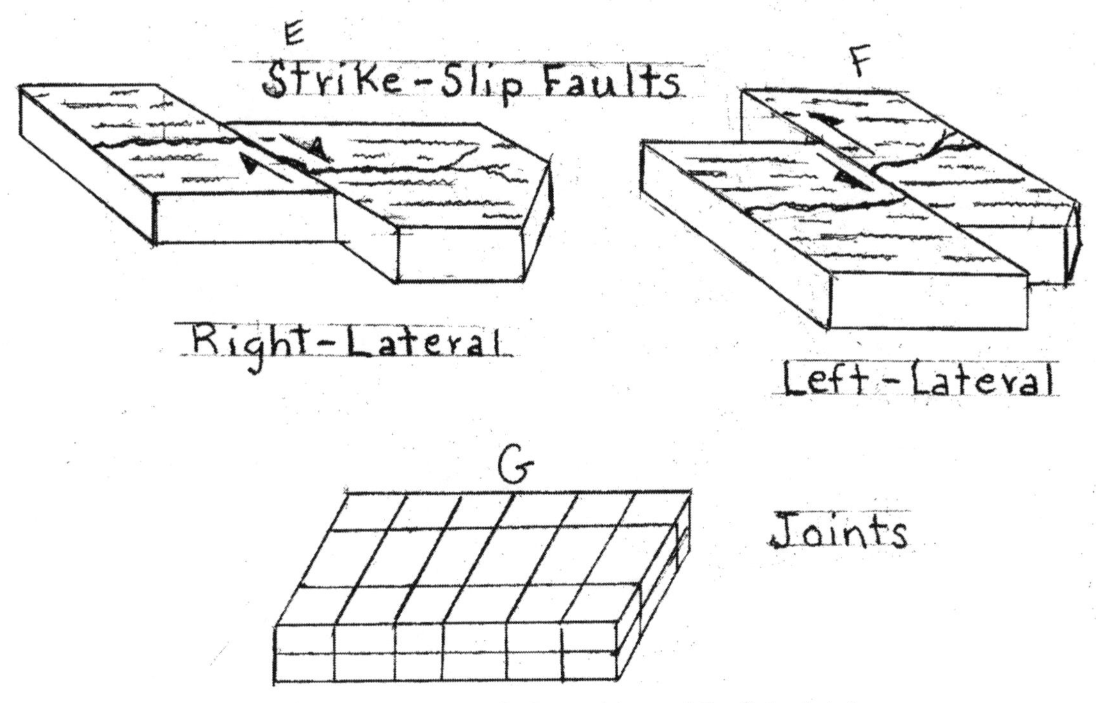


**Figure 13-7** A–G, From Normal Fault to Joints
*© A. Troell, 2008*

# NOTES

# CHAPTER 14

## Plate Tectonics

I. Continental Drift

  A. Pangaea

    1. Alfred Wegner proposed that the supercontinent of Pangaea was assembled about 250 million years ago

  B. Evidence used to support Wegener's continental drift hypothesis

    1. Fit of the continents

      a. Best fit is in about 6000 feet of water along the continental slope

    2. Matching fossils across oceans

      a. *Mesosaurus* (freshwater reptile) and *Glossopteris* flora (fossil plant assemblage) found on widely separated continents

    3. Matching rock types and mountain ranges

      a. Appalachian Mountains

        i. Extend along east coast of North America, disappear around Newfoundland

      b. Mountains of similar age and deformational style in North America, Greenland, British Isles, and Scandinavia

    4. Climate data

      a. Continental glaciation on the southern continents (220–300 million years ago)

      b. At the same time coral reefs and swamps in Northern Hemisphere

      c. Reassemble Pangaea with South Africa over the South Pole

        i. Direction of glacial movement makes sense and brings parts of northern continents nearer the tropics, so it is consistent with fossil evidence

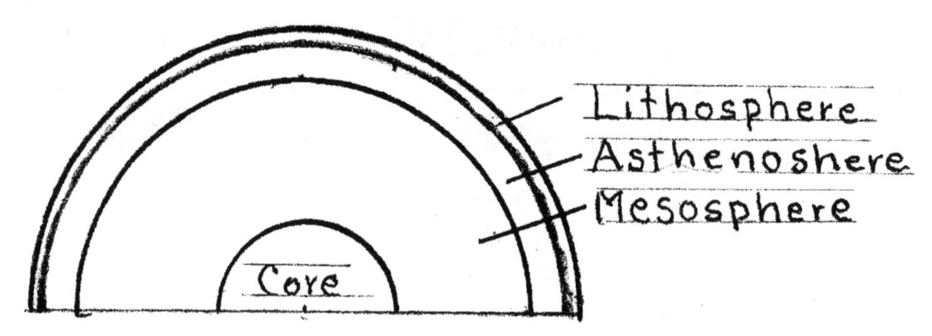


**Figure 14-1** Lithosphere, Asthenosphere, and Mesosphere
*© A. Troell, 2008*

II. Plate Tectonics

A. The theory

1. Lithosphere behaves as a brittle solid and rests on the hotter, weaker asthenosphere (uppermost mantle)–Fig. 14-1
2. The lithosphere is broken into plates–all of the large plates (except Pacific plate) consist of both oceanic and continental crust and uppermost brittle mantle
3. Lithospheric plates move slowly over the underlying asthenosphere (Fig. 14-2)
   a. Separate at oceanic ridges
   b. Collide at subduction zones
   c. Move past each other at transform boundaries
4. Types of lithosphere
   a. Oceanic lithosphere is overall "basaltic," denser and thinner than continental crust
   b. Continental lithosphere is overall "granitic," less dense and thicker than oceanic crust
5. Plates interact along their boundaries (edges)
6. One plate may have all three types of boundaries around its edges
7. Rates of plate movement are slow–average of 5 cm/yr

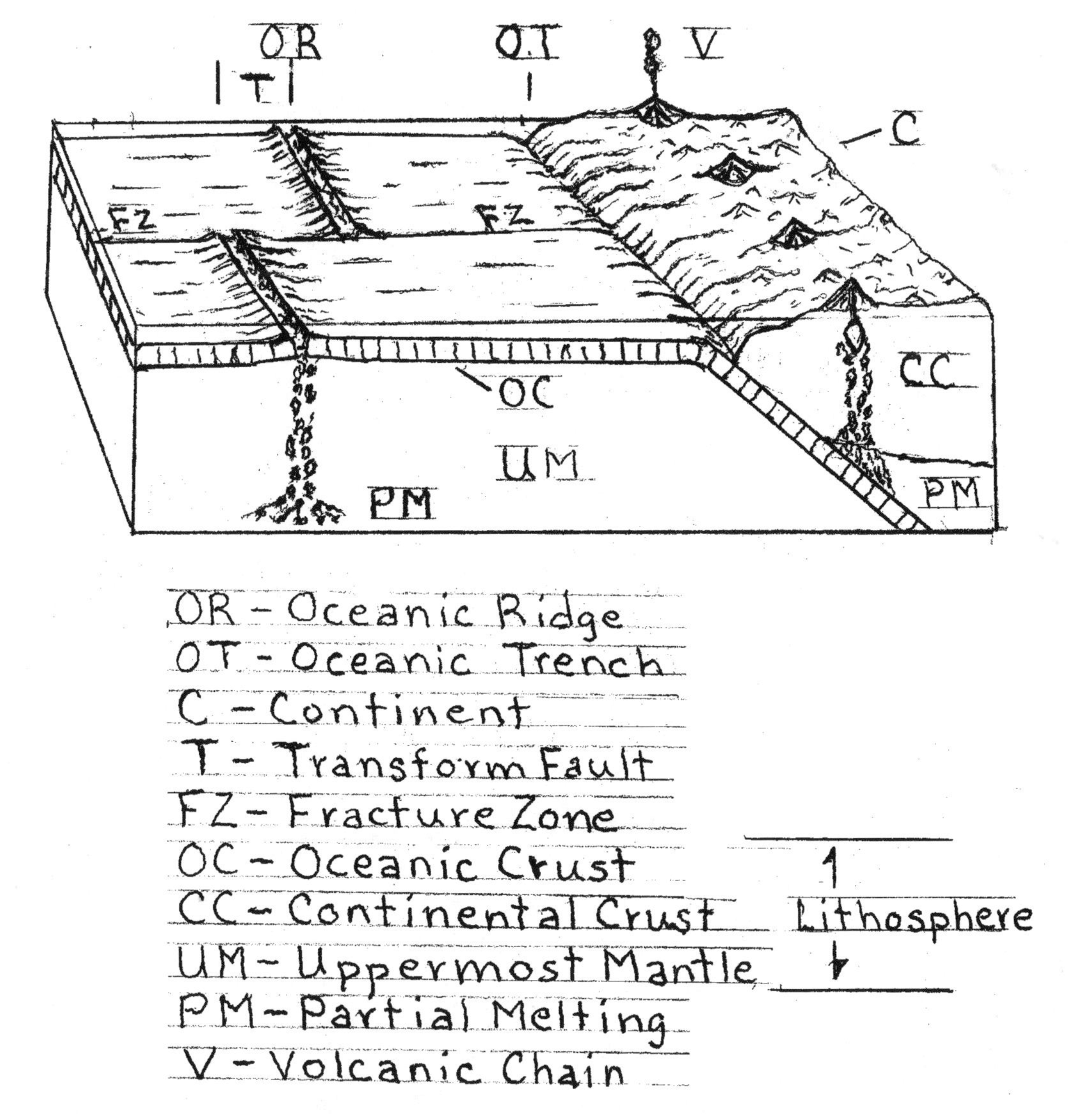


**Figure 14-2** From Oceanic Ridge to Volcanic Chain
*© A. Troell, 2008*

III. Plate Boundaries

A. Divergent–plates move apart

1. Oceanic ridges (Fig. 14-3D)

a. Form underwater mountain chain that stretches for 43,000 miles around world

b. Basalt cools to produce new oceanic lithosphere

c. Shallow focus earthquakes occur here

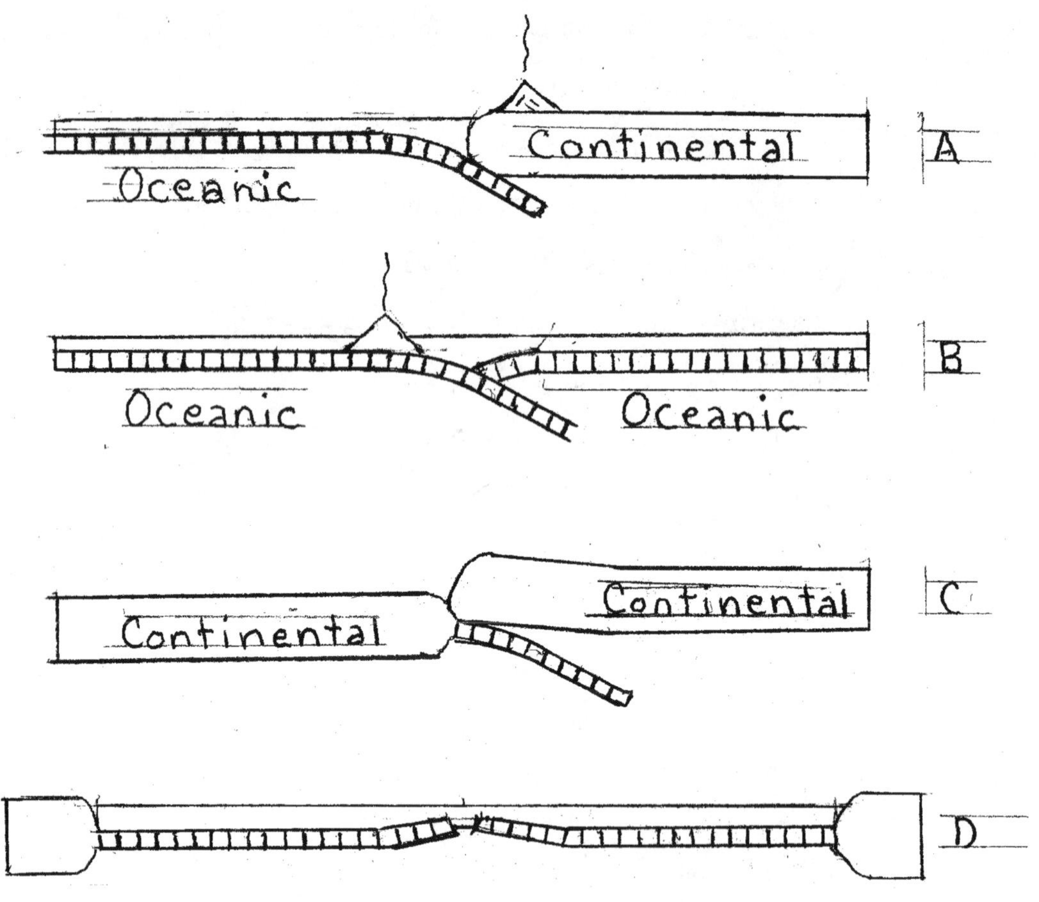


**Figure 14-3** A-C, Plate Convergence; D, Plate Divergence


2. Spreading centers under continents
    a. Continents break apart–molten rock rises from asthenosphere
    b. As plates move apart, rift valleys develop
    c. East African Rift Valley
3. Stages of formation of ocean basins
    a. Initial stage–East African Rift valley
    b. Later stage–Red Sea or Gulf of California (sea forms between continents)
    c. Late stage–ocean basin formed–Atlantic Ocean

B. Convergent plate boundaries–two plates collide
1. Oceanic-Continental plate boundary (Fig. 14-3A)
    a. Oceanic lithosphere subducts (descends) beneath continental plate
    b. Trench parallels volcanic arc
        i. Physical depression on the seafloor

c. Magma created as plate descends, heats up and water is driven off oceanic crust into overlying mantle
   i. Andesitic lavas erupt to form continental volcanic arc that parallels trench
   ii. Igneous intrusions form the roots of mountains
d. Earthquakes generated as plate subducts
e. Examples: Andes Mountains (South America) and Cascades (North America)

2. Oceanic-Oceanic plate boundary (Fig. 14-3B)
   a. Older, denser oceanic plate subducts
   b. Island arc is created–parallels the trench
      i. Andesitic volcanoes form on the seafloor
   c. Earthquakes generated
   d. Examples: Aleutian Islands & Tonga Islands
3. Continental-Continental plate boundary (Fig. 14-3C)
   a. When two continents converge, neither will subduct (both are too buoyant)
   b. Folded mountain belt
      i. Rocks have been faulted and folded
      ii. Igneous intrusions (plutons)
      iii. Metamorphic rocks
      iv. Pieces of oceanic crust
   c. Examples: Himalayas (Asia) and Appalachian Mountains (NA)–Fig. 14-4

C. Transform boundaries–where two plates move past each other
   1. Make relative motion of the plates possible
   2. Most transform faults connect segments of oceanic ridges
   3. Some transform faults cut through continental crust–San Andreas Fault connects the spreading ridge in the Gulf of California and the Cascadia subduction zone (Oregon and Washington)

IV. Evidence for Plate Tectonics
   A. Paleomagnetism
      1. Polar wandering–continents have moved in relation to a "fixed" pole
      2. Magnetic reversals–the earth's magnetic field switches polarity; the South Pole becomes the dominant pole during a time of reverse polarity

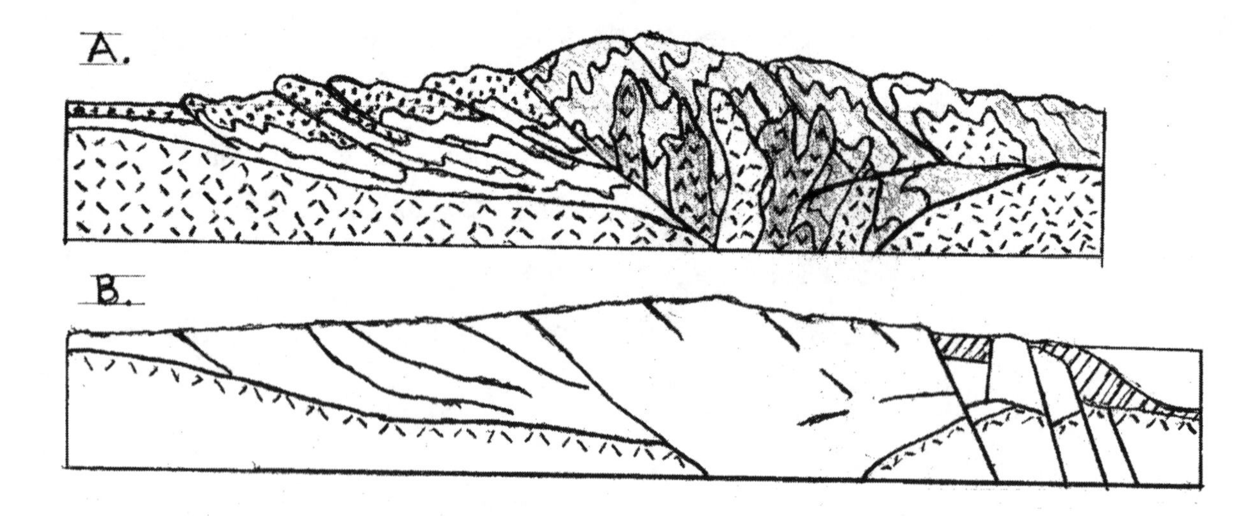


Section Across Appalachian Mountains
A. Late Paleozoic Orogeny
B. Early Mesozoic Rifting & Deposition

**Figure 14-4** Section Across Appalachian Mountains
*© After P. B. King*

B. Ocean drilling

   1. Indicates that there is no oceanic crust older than 180 my

   2. Oceanic crust increases in age with increasing distance from an oceanic ridge

C. Earthquake patterns

   1. Close association between earthquakes and plate boundaries

   2. Benioff zones–inclined zones of earthquakes trenches

D. Hot spots

   1. Volcanoes in the Hawaiian Island-Emperor Seamount chain increase in age in NW direction

   2. The only active volcanoes are on the island of Hawaii which is located directly over a hot spot

# NOTES

# NOTES

# CHAPTER 15

## Mountain Building

I. Mountains

  A. Areas that rise significantly above the surrounding landscape

  B. Orogeny–period of mountain building

    1. Compressional folding and thrusting

    2. Igneous activity

    3. Metamorphism

II. Subduction and Mountain Building

  A. Oceanic-oceanic plate boundaries (Fig. 15-1)

    1. Older, denser oceanic plate subducts

    2. Volcanic island arc

    3. Partial melting of mantle

      a. Mantle has low density and rises

      b. Andesitic lava and pyroclastics

    4. Can produce mountains with igneous and metamorphic rocks

    5. Aleutian Islands and Tonga Islands

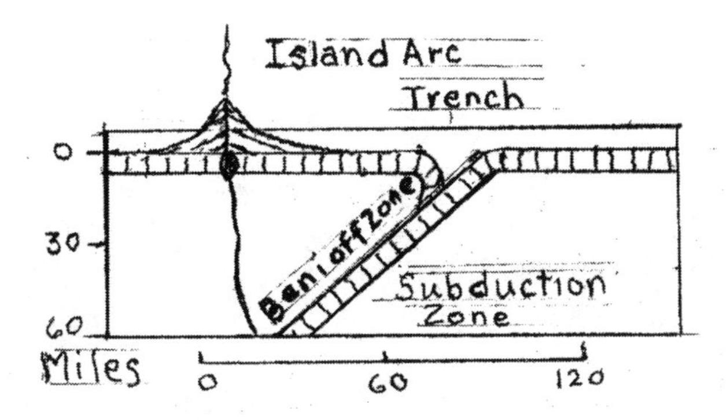


**Figure 15-1** Island Arc Trench
© A. Troell, 2008

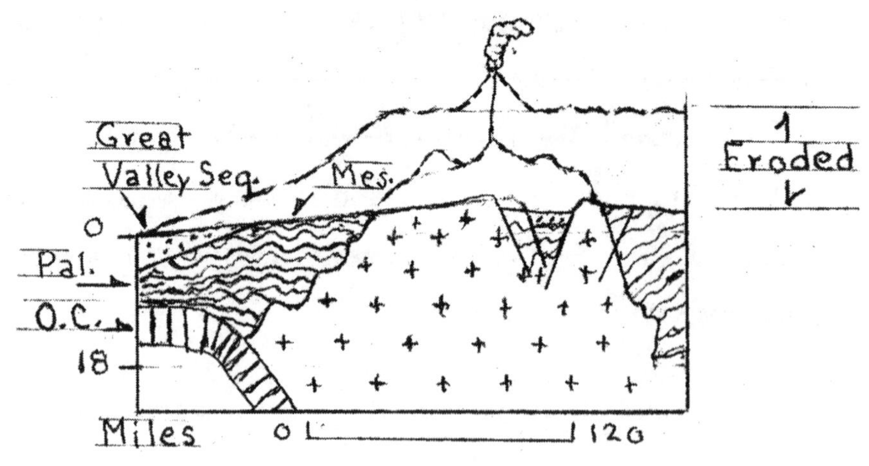


**Figure 15-2** Sierra Nevada Batholith
*© After P. B. King*

B. Continental-oceanic plate boundaries (Fig. 15-2)

1. Continental volcanic arc forms on a continent
2. Melting occurs along the subduction zone
    a. Magmas/lavas of andesitic to granitic composition
3. Thick continental crust slows ascent of magma producing massive plutons (igneous intrusions)
4. Andes Mountains of South America
    a. The west coast of South America started as a passive continental margin
        i. Have thick wedges of sedimentary rocks
        ii. Passive margins eventually become active margins
    b. The west coast of South America became an active continental margin when it collided with the Nazca plate–west coast of NA
        i. Have a subduction zone and continental volcanic arc
        ii. Accretionary wedge–formed of material (oceanic crust and deep-sea sediments) scraped off the subducting plate
5. Sierra Nevada of California (Fig. 15-2) - eroded volcanic arc
    a. The Sierra Nevada is an eroded continental volcanic arc
    b. The Franciscan sediments of the Coast Ranges (of California) are composed of accretionary wedge material

C. Continental-Continental Collisions (Folded Mountains)—See Plate Tectonics, Fig. 15-4

1. Form as a result of continental-continental collisions
    a. Subduction completed and collision occurs
2. Collision deforms margins of both continents with subsequent thickening
3. Includes folding, thrusting, emplacement of plutons (igneous intrusions) and metamorphism
4. Crust is shortened and thickened
5. Folded Mountains
    a. Himalayas
        i. India began colliding with Eurasia 40 my ago
    b. Appalachians (see Plate Tectonics, Fig. 15-4)
        i. Collision of North America, Europe, and Africa
        ii. Final orogeny 250–300 my ago
        iii. Subsequent erosion has lowered the mountain

III. Extension of Continents

A. Origin

1. Mantle upwelling and intrusion, accompanied by volcanism, domes, and extension of the crust
2. Erosion removes overburden and crustal uplift occurs by isostatic adjustment
    a. Crust of the earth sinks into the mantle similar to an ice cube floating in water
    b. As erosion occurs, the crust rises isostatically like the continuous melting of an ice cube
    c. By this process, the "roots" of mountains may be exposed at Earth's surface

B. Fault-block mountains

1. Mountains that are bounded on at least one side by moderate to high-angle normal faults
2. Teton Range, Wyoming (Fig. 15-3)
3. Basin and Range Province (southern Oregon to Mexico)
4. Sierra Nevada of California

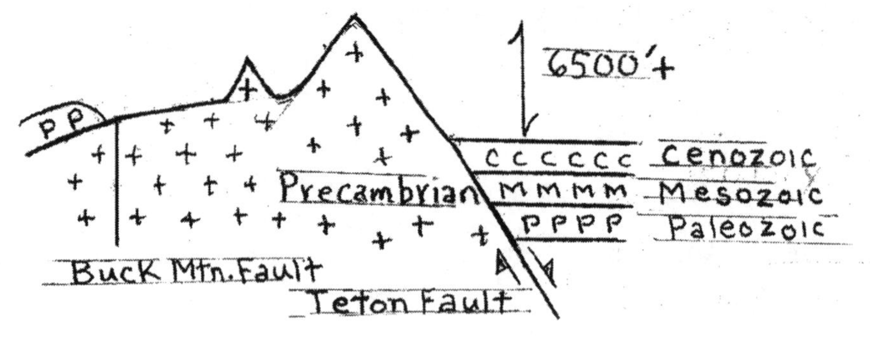


**Figure 15-3** Teton Range, Wyoming

*© A. Troell, 2008*

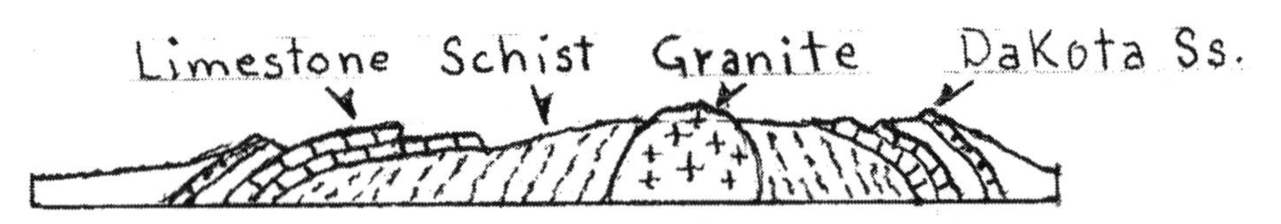


**Figure 15-4** Black Hills, South Dakota

*© After A. N. Strahler*

C. Upwarped mountains

1. Mountains formed as a result of broad arching of the crust
2. Black Hills, South Dakota (Fig. 15-4)

IV. Terranes

A. Collison of continents with small crustal fragments may cause an orogeny

B. Types of crustal fragments

1. Volcanic island arcs
2. Microcontinents

principle of isosty

- less dense crust "floats" in equilimbrium w/ denser underlying mantle

- As mountains erode, their roots rise to surface

# NOTES

# NOTES

# CHAPTER 16

## Earthquakes

I. Earthquakes

   A. Sudden vibration of the earth caused by the release of energy

   B. Most earthquakes are generated by movement along a fault

   C. Focus–area inside the earth where the energy is released (Fig. 16-1)

      1. Shallow focus earthquakes–occur less than 40 miles within the earth

      2. Deep focus earthquakes–occur greater than 180 miles within the earth

   D. Epicenter–location on Earth's surface directly above the focus (Fig. 16-1)

   E. Foreshocks–smaller earthquakes that precede a larger earthquake

   F. Aftershocks–smaller earthquakes that follow a large earthquake

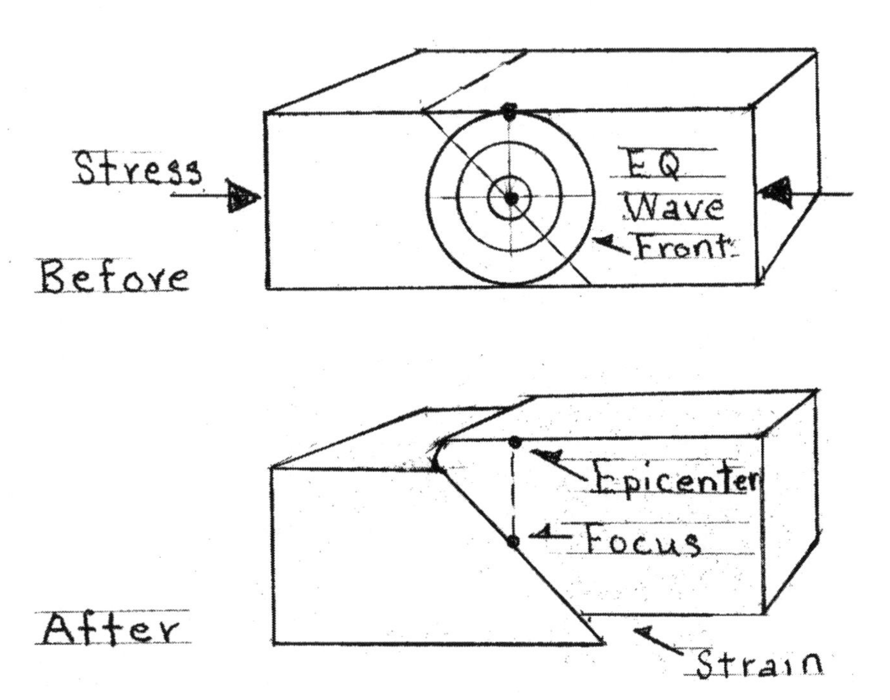


**Figure 16-1** Occurrence of an Earthquake

G. Elastic Rebound Theory

1. Rocks along an active fault plane stick
2. Rocks on either side of fault slowly bend and store energy
3. Eventually, internal strength of rocks along fault plane are exceeded, and they rupture
4. Rocks on either side of fault plane "snap back" to their original position, releasing energy that radiates in all directions

II. Seismology

A. Study of earthquake waves

B. Types of seismographs (Fig. 16-2)

1. Vertical motion detector
2. Horizontal motion detector

C. Seismogram—seismograph record (Fig. 16-3)

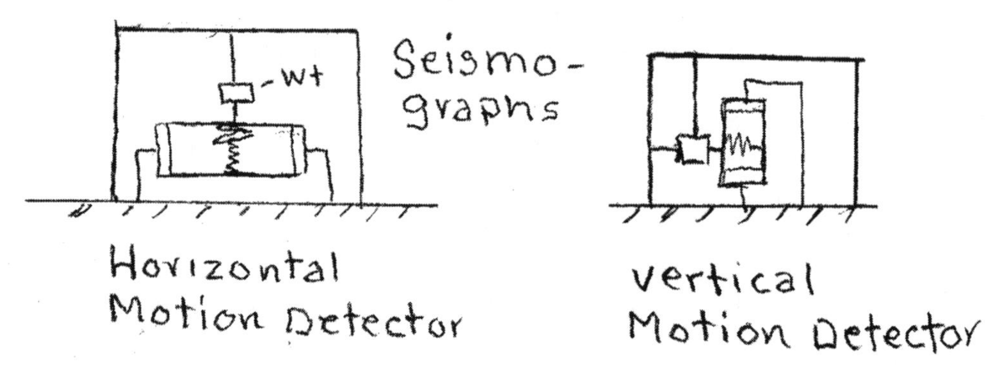


**Figure 16-2** Seismographs—Horizontal and Vertical



Horizontal
Motion Detector

vertical
Motion Detector

First P
First S
Surface Waves
Seismogram
Time

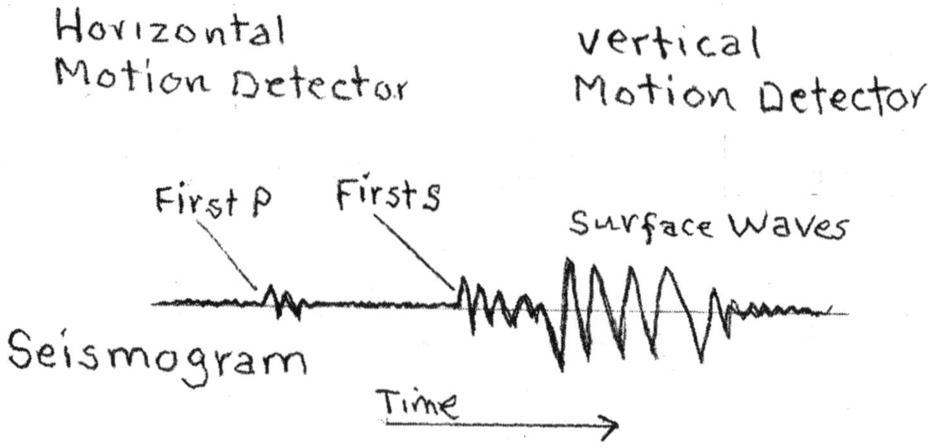

**Figure 16-3** Seismogram

III. Earthquake Waves

A. Body waves–move through the Earth's interior (Fig. 16-4)

1. P-waves (Primary waves)

a. Arrive first at seismic recording station

b. Rocks expand and contract as waves move through them

c. Travel through solids, liquids, and gases

2. S-waves secondary

a. Arrive second at a recording station

b. Shake rocks at right angles to their direction of travel

c. Travel only though solids

B. Surface waves (Long or L-waves)–travel along Earth's surface

1. Cause much of the damage associated with earthquakes

IV. Locating Earthquake Epicenters (Fig. 16-5)

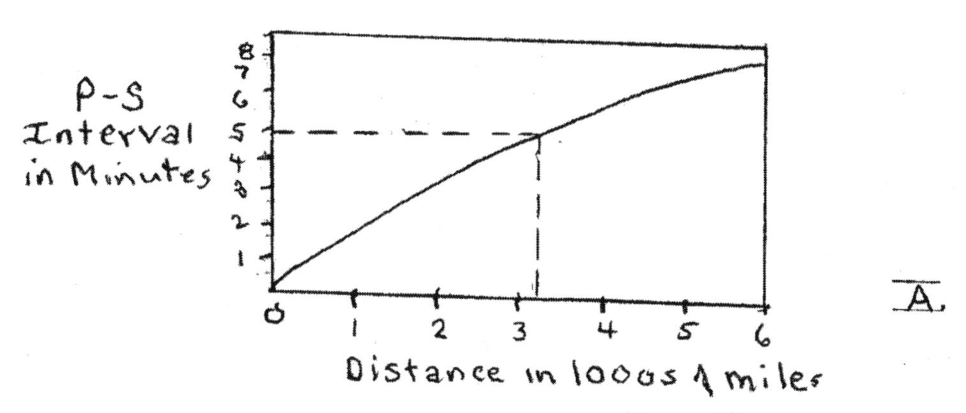


**Figure 16-4** P-S Interval in Minutes
*© A. Troell, 2008*

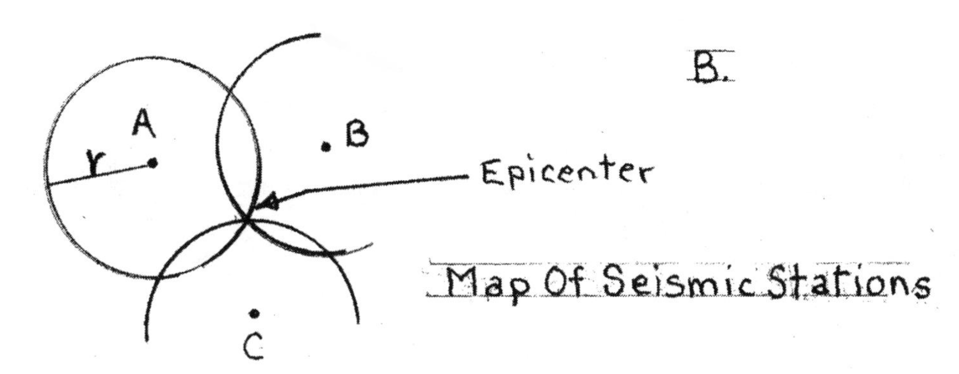


**Figure 16-5** Map of Seismic Stations
*© A. Troell, 2008*

V. Measuring Earthquakes

A. Modified Mercalli intensity scale

1. Measures damage done and people's reactions to earthquakes

2. Used primarily by insurance companies

B. Richter scale

1. Magnitude scale that measures amplitude of the largest wave recorded

2. The difference between any two numbers represents a 32-fold increase in energy released

VI. Damages Associated with Earthquakes

A. Ground shaking–causes the most damage and loss of life

1. Geology and ground shaking

a. Solid bedrock–structures built on solid bedrock generally suffer less damage

b. Liquefaction–water-saturated sediment may lose cohesion when shaken and flow

2. Building materials and ground shaking

a. Wood and reinforced concrete tend to be more flexible

b. Brick and unreinforced concrete tend to crumble

B. Tsunami

1. Series of waves generated by earthquakes, volcanic eruptions, or underwater landslides

2. Are 3–6 feet high when they travel across the ocean

3. As the waves enter shallow water, they slow and begin to build

4. Tsunami may come onshore as waves or as a rapid rise in water

C. Landslides–vibrations from earthquakes may generate mass wasting events

D. Ground subsidence–the ground surface may subside due to movement along faults

E. Fire

# NOTES

# NOTES

# CHAPTER 17

## Earth's Interior

I. Introduction

   A. Study the movement of earthquake (body) waves through earth's interior

   B. Paths of seismic waves are bent (refracted) as they move

   C. Abrupt changes in velocities at discontinuities (boundaries)

II. Layers Defined by Composition (Fig. 17-1)

   A. Crust

      1. Oceanic crust

         a. 3–6 miles thick

         b. Density = 3.0 g/cm$^3$

      2. Continental

         a. 12–50 miles thick

         b. Density = 2.7 g/cm$^3$

   B. Mantle

      1. Extends to 1800 miles

      2. Composed of peridotite

      3. Average density = 4.5 g/cm$^3$

   C. "Moho"–Crust-mantle boundary (Fig. 17-2)

   D. Core

      1. Iron-nickel alloy

      2. Radius–2150 miles

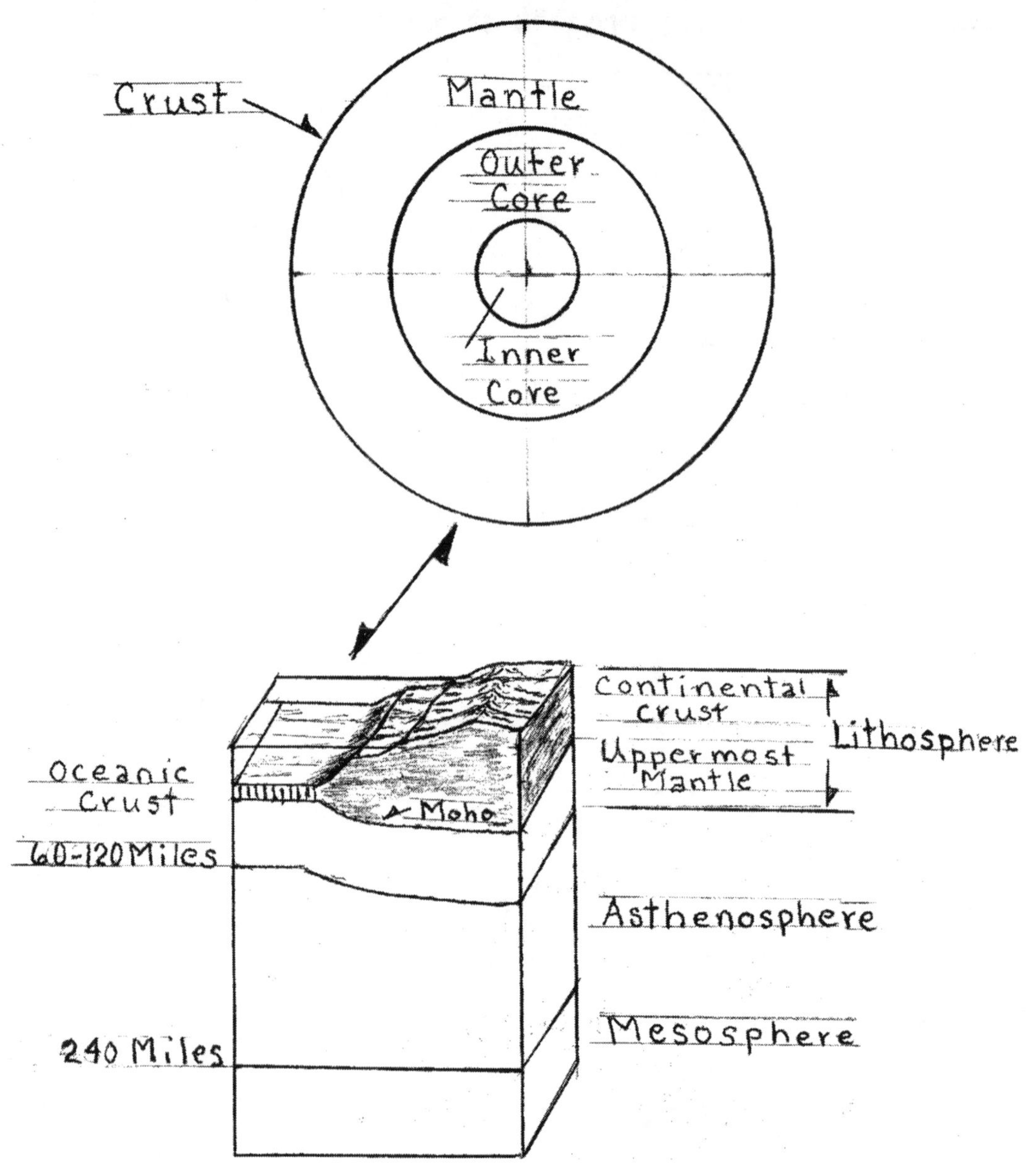


**Figure 17-1** Layers of the Earth–Crust, Mantle, Outer Core, Inner Core
**Figure 17-2** More Layers–Oceanic Crust, Continental Crust, etc.
*© A. Troell, 2008*

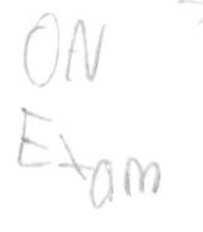


III. Layers Defined by Physical Properties (Fig. 17-2)

A. Lithosphere (crust and uppermost brittle mantle)

1. Cooler, rigid layer
2. Averages 60 miles in thickness, but may be up to 150+ miles below older parts of continents

B. Asthenosphere (low velocity zone of the mantle)

   1. Soft, weak layer

C. Mesosphere

   1. Lower mantle (solid)

   2. Rocks hot and capable of very gradual flow

D. Outer Core

   1. Behaves as a liquid

   2. 1350 miles thick

   3. Average density = 11 g/cm$^3$

   4. Convective flow of iron within zone generates Earth's magnetic field

E. Inner core

   1. Behaves as a solid

   2. Radius 750 miles

   3. Average density = 12.8 g/cm$^3$

   4. Rotates faster than rest of Earth

IV. Discovering Earth's Major Layers

A. Crust/mantle boundary (1909)–Velocity of P-waves increases abruptly below 30 miles

B. Core-mantle boundary

   1. Located at about 2900 km

   2. P-wave shadow zone (1914)

      a. Area where direct P-waves are absent at seismic stations between 105–140 degrees from an EQ

      b. P-waves are slowed and bent, so that few P-waves occur in the shadow zone

C. Liquid behavior of outer core (1926)–S-waves are completely blocked by core creating an S-wave shadow zone (105 to 256 degrees) from an EQ

D. Existence of inner core (1936)–The weak P-waves observed in the P-wave shadow zone are refracted as they move from outer core to the inner core

# NOTES

# CHAPTER 18

## Volcanoes and Plutons

I. Explosivity of Volcanic Eruptions

   A. Viscosity–resistance to flow of lava or magma

      1. Silica content influences viscosity

         a. The higher the silica content, the more viscous the magma/lava

            i. Basaltic lavas–50% silica

            ii. Andestitic lavas–60% silica

            iii. Rhyolitic lavas–70% silica

      2. Temperature–the higher temperature, the less viscous the lava

   B. Importance of dissolved gases

         a. Gases are held in magma by confining pressure

         b. As magma moves up, the confining pressure is reduced, and dissolved gases expand

   C. Basaltic eruptions tend to be less violent–less viscous lava, hotter, lower gas content

   D. Rhyolitic eruptions tend to be more violent–more viscous lava, cooler, higher gas content

II. What Is Erupted During Volcanic Eruptions

   A. Lava flows

      1. Pahoehoe flows–fluid, ropy basaltic flows

      2. aa flows–surface of rough, jagged blocks, tend to be cool and thick

      3. Pahoehoe lava flows may become aa flows as they cool

   B. Gases

      1. Water vapor–most abundant gas erupted

      2. Carbon dioxide–second most abundant gas erupted

      3. Eruption of volcanic gases formed earth's atmosphere and hydrosphere

C. Pyroclastic materials

1. Fragments ejected explosively from a volcano

2. May range in size from dust and ash to block, bombs

III. Types of Volcanoes

A. Features of volcanoes

1. Vent–opening in the crust through which material is ejected

2. Pipe–feeds magma from the magma chamber to the surface

3. Crater–steep-walled depression at the summit of a volcano, built through successive eruptions

4. Caldera–large, circular depressions formed through collapse of the summit

5. Fumarole–vents that erupt only gases

B. Types of volcanoes

1. Shield volcanoes (Fig. 18-1)

a. Built by successive basaltic lava flows

b. Broad, slightly domed shape

c. Hawaiian Islands, Galapagos Islands

2. Cinder cones (Fig. 18-2)

a. Built primarily by successive eruptions of pyroclastic materials

b. Small, steep-sided, usually short-lived

c. May form parasitic cones on larger (composite cones)

d. Paricutin, Mexico

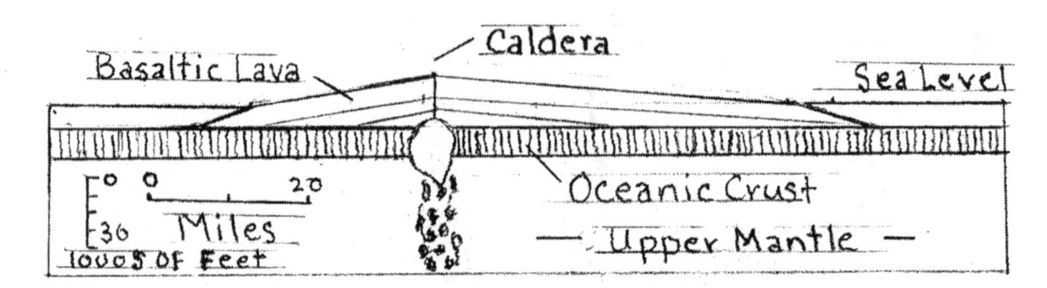


**Figure 18-1** Cross Section of Mauna Loa Shield Volcano, Hawaii

3. Composite cones (Figs. 18-3 and 18-4)

    a. Built by alternating eruptions of andesitic lava flows and pyroclastic materials

    b. Large, symmetrical cones

    c. Cascade Range, Andes Mountains

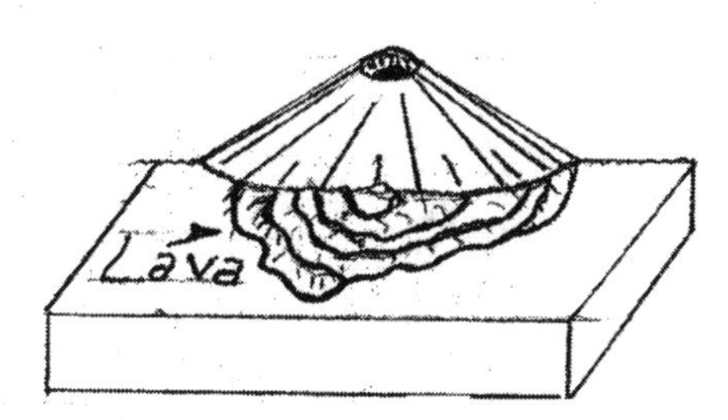


**Figure 18-2** Cinder Cone

*© A. Troell, 2008*

Crater

Mount Shasta, N.Cal.

Composite Volcano

Andesitic Composition

15 Miles Wide

11,050 Fee High

**Figure 18-3** Crater—Mount Shasta, North Carolina

*© A. Troell, 2008*

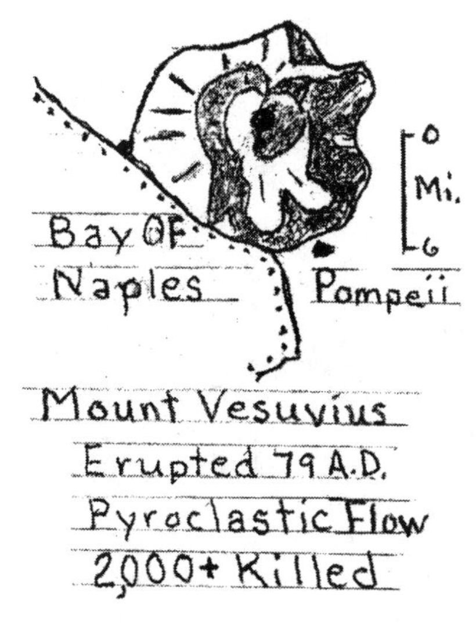


**Figure 18-4** Mount Vesuvius

*© A. Troell, 2008*

C. Volcanic hazards associated with composite cones

1. Nuee ardente

a. Hot, fiery cloud of gas and ash, often created by collapse of a lava dome

b. Mt. Pelee, Martinique

2. Lahar

a. Volcanic mudflows

b. Mt. St. Helens (1980), potential on Mt. Rainier for lahars

IV. Other Volcanic Landforms

A. Calderas

1. Large, collapsed depressions with a diameter greater than 1 km

2. Crater Lake-type calderas

a. Formed when Mt. Mazama (composite cone) violently erupted pyroclastic material and the cone suddenly collapsed

b. The caldera eventually filled with water

c. Crater Lake, Oregon

3. Hawaiian-type calderas

a. Formed by gradual subsidence of the summit as magma slowly drained from magma chamber to a rift zone

4. Yellowstone-type calderas

a. Formed when rhyolitic magma is emplaced near surface

b. Fractures develop and provide a pathway to surface for gas-charged magma, which causes a very explosive eruption

c. Yellowstone National Park is located in a caldera

B. Fissure eruptions and lava plateaus

1. Very fluid basalt erupts from fissures (fractures in the crust)

2. Fissure eruptions create lava plateaus

3. Columbia River basalts created the Columbia River Plateau (Oregon and Washington)

- C. Volcanic pipes and necks
  1. Volcanic pipes–connect magma chamber to the surface
  2. Volcanic necks–eroded remnants of volcanic pipes
     - a. Shiprock, NM

V. Intrusive Igneous Activity

- A. Plutons–intrusive igneous bodies – Magma inside Rock
- B. Classifying plutons
  1. Shape–tabular (tabletop) or massive (no shape)
  2. Orientation to surrounding rocks
     - a. Discordant–cut across existing rocks
     - b. Concordant–intrude between rock layers = Parallel
- C. Types of plutons (Fig. 18-5)
  1. Dikes
     - a. Tabular, discordant
     - b. Often form when magma fills fractures
  2. Sills–tabular, concordant
  3. Laccoliths
     - a. Mushroom-shaped, concordant
     - b. Generally form at shallow depths
  4. Batholiths → Coarse grained / largest types
     - a. Massive, discordant
     - b. Greater than 40 square miles in surface extent

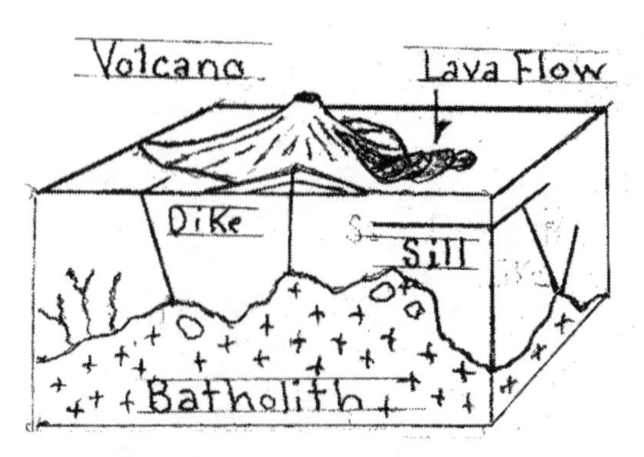


**Figure 18-5** Igneous Rock Bodies
*© A. Troell, 2008*

# NOTES

# CHAPTER 19

## Geologic Time

I. Birth of Modern Geology

  A. James Hutton

  B. Uniformitarianism–"The present is the key to the past"

II. Relative Dating (Placing events in order of occurrence)–Key Principles

  A. Law of Superposition (Fig. 19-1A)

    1. In an undeformed sequence of sedimentary rocks, each bed is older than the one above it and younger than the one below it

  B. Principle of Original Horizontality (Fig. 19-1A)

    1. Layers of sediment are deposited in a horizontal position

    2. If rocks are folded or steeply dipping (inclined), they must have been moved to that position after deposition

  C. Principle of Cross-Cutting Relationships–A fault or igneous intrusion that cuts across rock layers is younger than the rocks it disrupts (Fig. 19-2)

  D. Inclusions

    1. Pieces of one rock that are contained within another rock

    2. Inclusions are always older than the rock they are contained in

      a. Xenoliths–inclusions in igneous rocks (Fig. 19-1C)

      b. Clasts–inclusions in sedimentary rocks (Fig. 19-1D)

  E. Unconformities

    1. Conformable sequence–sequence of rocks that have been deposited without interruption

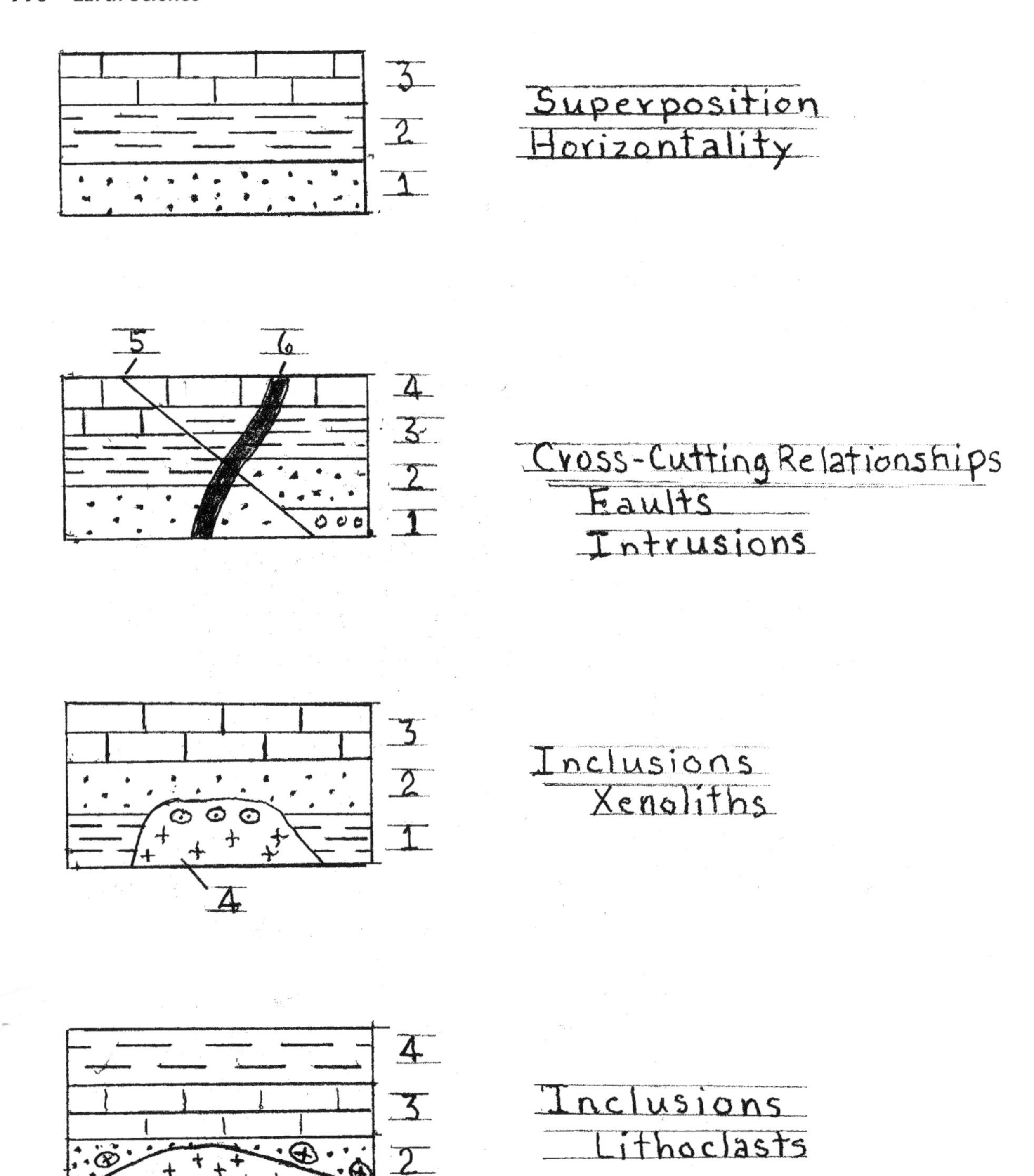


**Figure 19-1** Principles of Relative Geologic Time
*© A. Troell, 2008*

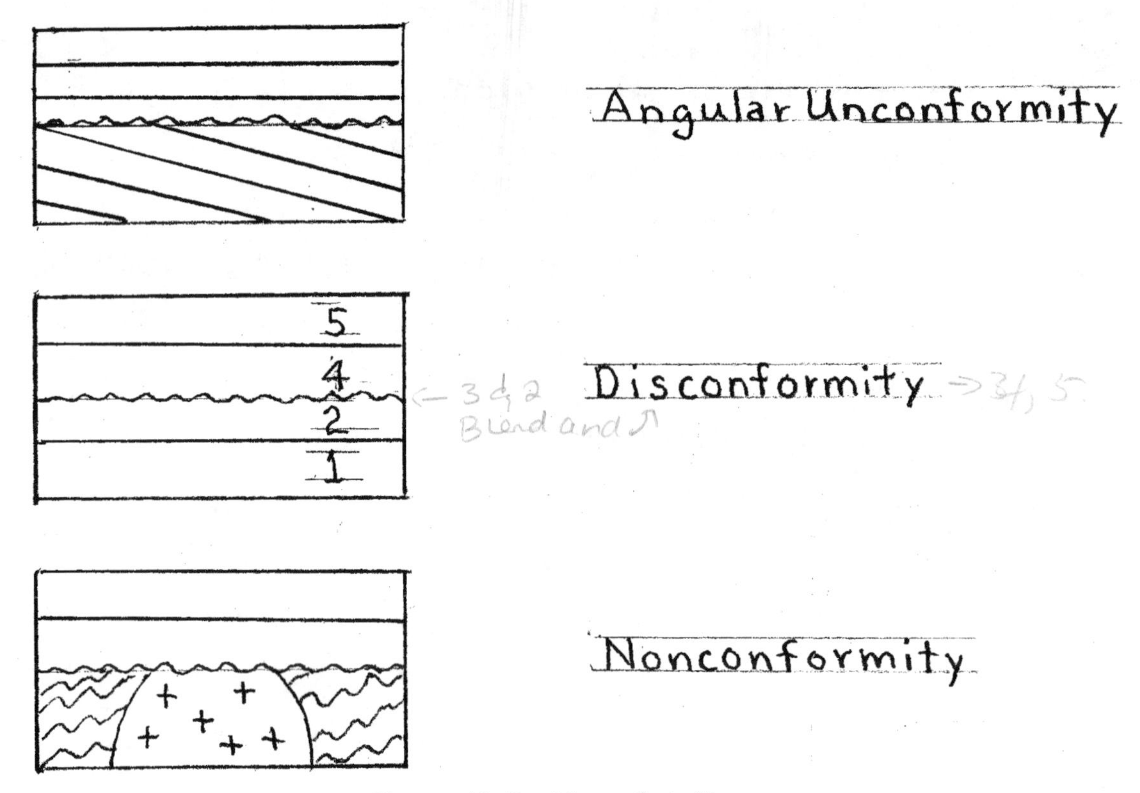


**Figure 19-2** Unconformities
*© A. Troell, 2008*

2. Unconformities
   a. Buried surfaces of erosion or nondeposition
   b. Time during which deposition stopped, erosion occurred, then deposition resumed
3. Types of unconformities
   a. Angular unconformity–Erosional surface developed on tilted or folded rocks that is overlain by younger, more flat-lying strata (Fig. 19-2A)
   b. Disconformity–Strata on either side of the unconformity are essentially parallel (Fig. 19-2B)
   c. Nonconformity–Unconformity that separates underlying igneous or metamorphic rocks from overlying younger sedimentary rocks (Fig. 19-2C)

* Earth is 4.5 Billion yrs old!

F. Principle of Fossil Succession

1. Fossils–remains of ancient animals and plants
2. Types of fossils and fossilization
   a. Body fossils
      i. Petrification–small internal cavities are filled with precipitated minerals
      ii. Mold–shell is buried, then dissolved by groundwater
      iii. Cast–mold is filled in with mineral matter
   b. Trace fossils–tracks, trails, burrows, etc.
3. Conditions favoring preservation
   a. Rapid burial
   b. Preservable hard parts

4. Principle of fossil succession
   a. Fossil organisms succeed one another in a definite order and any time period may be recognized by its fossils
   b. Fossils document the evolution of life on earth
   c. Index fossils
      i. Widespread geographically
      ii. Lived for a short interval of geologic time
      iii. Easily recognizable

III. Absolute Geologic Time

A. Parent and daughter products of radioactive decay

1. Parent–unstable isotope
2. Daughter product–result from radioactive decay

B. Half Life–time it takes in years for one-half of parent to decay to the daughter product (Fig. 19-3)

C. Examples of parent, daughter, and half-life

1. $C^{14} => N^{14}$ 5,730 +/– 30 years
2. $U^{235} => Pb^{207}$ 713 my
3. $U^{238} => Pb^{206}$ 4.56 by

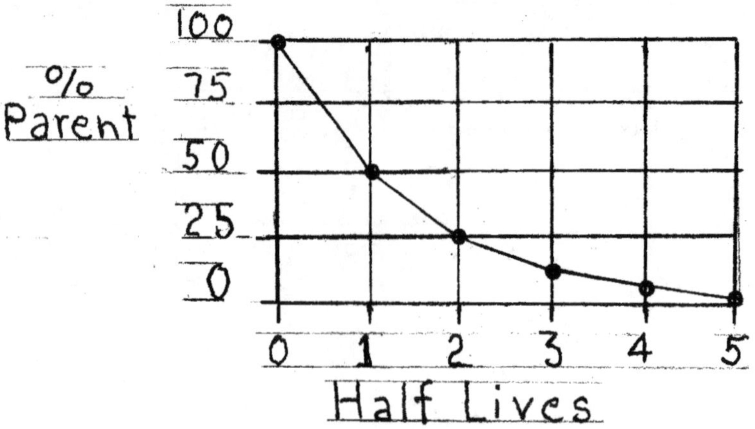

**Figure 19-3** Half Lives
*© A. Troell, 2008*

IV. Correlation of Rock Layers

   A. Matching rocks of similar age in different regions

   B. Over limited areas, rocks may be matched based on rock type, position in a sequence of rocks, etc.

   C. Over long distances, fossils are used for correlation

V. The Geologic Time Scale (Fig. 19-4)

   A. Structure of the time scale

      1. Eons–largest time units

         a. Precambrian–represents about 88% of geologic time (4.5 by–545 my)

         b. Phanerozoic Eon–"visible life"

      2. Eras–eons are divided into eras

         a. Phanerozoic Eon

            i. Paleozoic Era–"ancient life" (545–245 my)

            ii. Mesozoic Era–"middle life" (245–66 my)

            iii. Cenozoic Era–"recent life" (66 my–present)

      3. Periods–eras are divided into periods

      4. Epochs–periods are divided into epochs

         b. Eons

            i. Eras

            ii. Periods

            iii. Epochs

| EON | ERA | PERIOD |
|---|---|---|
| 2. PHANEROZOIC | 5. CENOZOIC | 17. QUATERNARY |
| | | 16. TERTIARY |
| | 4. MESOZOIC | 15. CRETACEOUS |
| | | 14. JURASSIC |
| | | 13. TRIASSIC |
| | 3. PALEOZOIC | 12. PERMIAN |
| | | 11. PENNSYLVANIAN |
| | | 10. MISSISSIPPIAN |
| | | 9. DEVONIAN |
| | | 8. SILURIAN |
| | | 7. ORDOVICIAN |
| | | 6. CAMBRIAN |
| 1. PRECAMBRIAN | | |

**Figure 19-4** Different Eons, Eras, and Periods

*© A. Troell, 2008*

# NOTES

# NOTES

# CHAPTER 20

## Important Events in Life History

I. Precambrian

  A. First life appeared at 3.5 by

    1. Simple, single-celled bacteria and algae

  B. Late Precambrian

    1. Earliest multicelled organisms

II. Paleozoic Era

  A. Cambrian Period

    1. Earliest organisms with shells

    2. Earliest fish → give Rise to amphibians → Frogs, Toads etc

  B. Late Ordovician Period

    1. Earliest land plants

  C. Late Devonian Period

    1. Earliest amphibians

  D. Pennsylvanian Period

    1. Earliest reptiles—laid shelled eggs

  E. End of Paleozoic Era

    1. Largest mass extinction in earth history (on tests)

    2. ~90% of all marine species extinct

III. Mesozoic Era

A. Late Triassic Period

1. Earliest true mammals

2. Earliest dinosaurs

B. Late Jurassic Period

dinosaurs gave rise to birds.

1. Earliest birds

C. End of Mesozoic Era

1. Mass extinctions – dinosaurs & stuff

IV. Cenozoic Era

A. Age of mammals

# NOTES

# NOTES

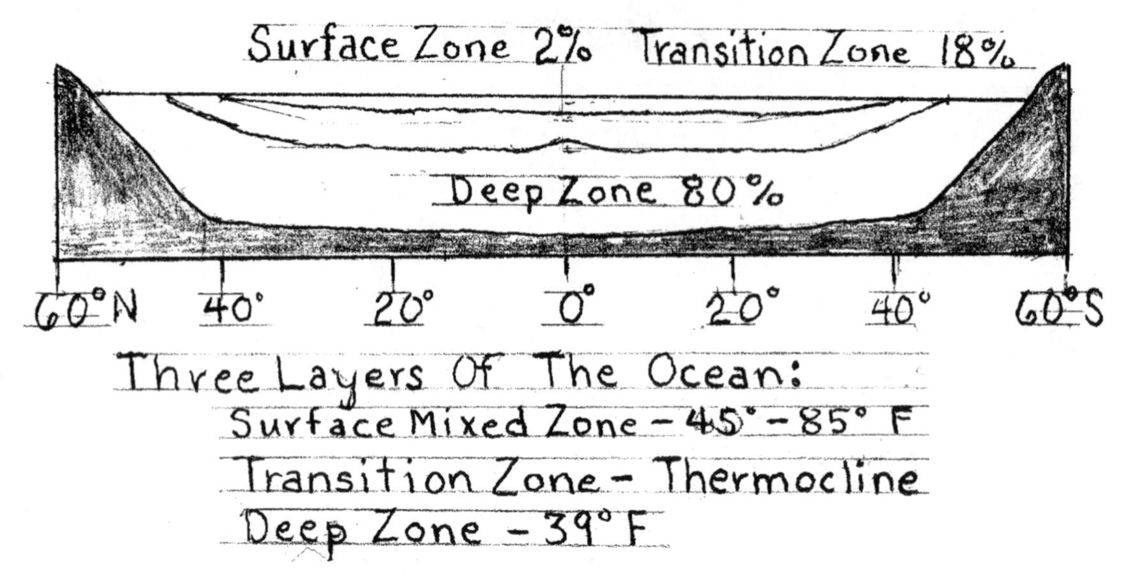


**Figure 21-1** Three Layers of the Ocean
*© A. Troell, 2008*

III. Oceans Layered Structure (Fig. 21-1)

A. Surface mixed zone–2% of volume of ocean

B. Transition zone–18% of volume of ocean

1. Thermocline–warm at top and colder at bottom

C. Deep zone–80%

1. Cold–39 degrees (F)

2. Originates at poles and sinks

IV. Lifestyles of Marine Organisms

A. Planktonic–floating

B. Nektonic–swimming

C. Benthonic–bottom dwelling

1. Epifaunal–live on ocean floor

a. Sessile benthonic–attached (can't move over seafloor)

b. Mobile benthonic–move over seafloor

2. Infaunal–burrow into ocean floor

# CHAPTER 21

## Oceans

I. Geography of the Oceans

  A. Overall

    1. Oceans–cover 71% of Earth's surface, mostly in Northern Hemisphere

    2. Land–covers 29% of Earth's surface

  B. Subdivisions of the oceans

    1. Pacific Ocean–50% of 71%

    2. Atlantic Ocean–25% of 71%

    3. Indian Ocean–21% of 71%

    4. Arctic Ocean–4% of 71%

II. Seawater

  A. Composition–ions in order of decreasing abundance

    1. $Cl^-$

    2. $Na^+$

    3. $SO^4$

    4. $Mg^+$

    5. $Ca^+$

    6. $K^+$

    7. $HCO_3^-$

  B. Sources of salt

    1. Chemical weathering of rocks on continents

    2. Volcanoes

  C. Changes in salinity occur

    1. Precipitation

    2. Evaporation

    3. Runoff from land

# NOTES

# NOTES

# CHAPTER 22

## Ocean Circulation

I. Deep Ocean Currents

  A. Cold, dense seawater originates at the poles by formation of ice and expulsion of salt from crystal structure of ice

  B. Cold water plunges near poles and moves along bottom driven by density

II. Surface Currents (Fig. 22-1)

  A. Driven by wind–move in gyres (large circles)

    1. Clockwise in Northern Hemisphere

    2. Counterclockwise in Southern Hemisphere

  B. Transfer head of solar origin form equator to poles

III. Coastal Upwelling

  A. Cold, nutrient-rich deep ocean water moves to surface along west coasts of continents

  B. Accounts for significant portion of fish food change

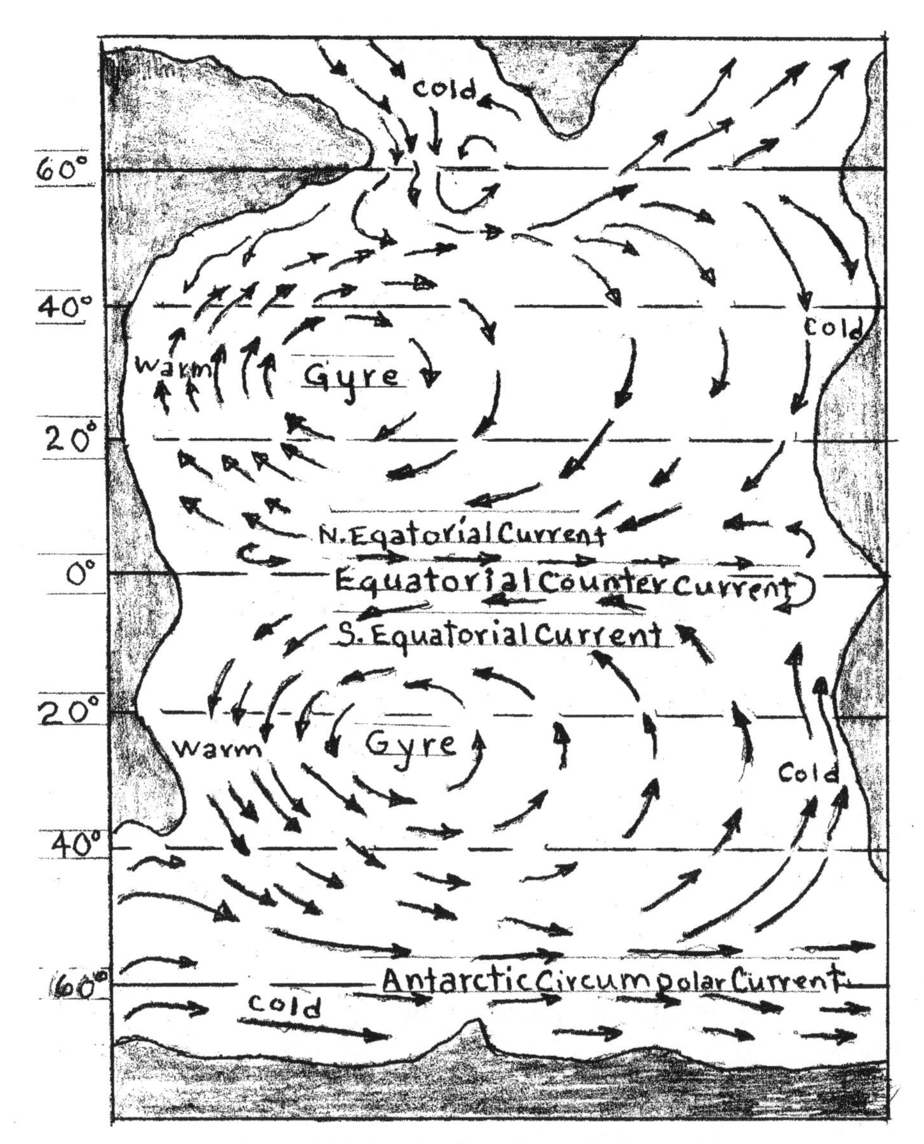


**Figure 22-1** Surface Oceanic Currents

*© After A. N. Strahler*

# NOTES

# NOTES

# CHAPTER 23

## Ocean Floor

I. Continental Margins

  A. Passive continental margins

    1. Edge of continent not associated with a plate boundary

    2. Primarily deposition of sediment derived from erosion of continents

  B. Features of passive continental margins (Fig. 23-1)

    1. Continental shelf

      a. Edge of the continent covered by ocean

      b. Slopes gently toward the ocean basin

      c. Up to 60 miles wide and 450–500 feet deep (at shelf edge)

    2. Continental slope

      a. Narrow feature that extends from shelf edge

      b. Inclined more steeply than continental shelf

      c. May be cut by submarine canyons

        i. Deep, steep-sided canyons eroded by turbidity currents

      d. Turbidity currents

        i. Dense mixtures of sediment and water that flow from continental shelf

        ii. When they reach the bottom of the continental slope, they slow and deposition occurs and deep-sea fans form

    3. Continental rise–Apron of sediment at the base of the continental slope composed of overlapping deep-sea fans

  C. Features of active continental margins (Fig. 23-1)

    1. Narrow continental shelf

    2. Earthquakes and volcanic activity

    3. May be associated with a trench

    4. Continental slope descends into a trench

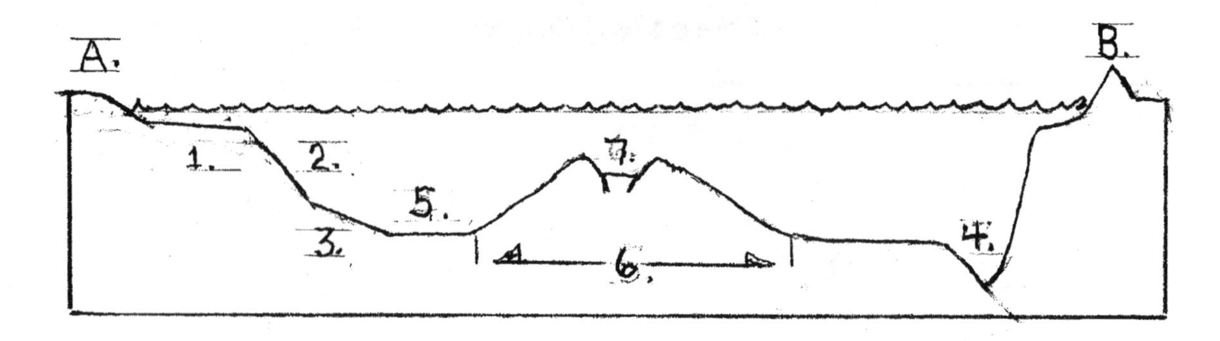


Types Of Continental Margins

A. Passive
1. Continental Shelf
2. Continental Slope
3. Continental Rise

B. Active
4. Oceanic Trench

Ocean Basin
5. Abyssal Plain
6. Oceanic Ridge
7. Rift Valley

**Figure 23-1** Types of Continental Margins
*© A. Troell, 2008*

II. Ocean Basins

A. Underlain by oceanic lithosphere

B. Trenches (Fig. 23-1)

1. Long, narrow, deep features formed where one plate is subducting beneath another plate

2. Associated with earthquakes, low heat flow, and andesitic volcanoes

C. Seamounts

1. Extinct submerged volcanoes

2. May form at divergent plate boundaries or over hot spots

D. Guyots

1. Submerged, flat-topped seamounts

2. Volcanoes that were above sea level, then became extinct and were eroded by wave action

E. Abyssal Plains (Fig. 23-1)

1. Flat parts of the ocean where oceanic lithosphere is covered by thick deposits of seafloor sediment

III. Mid-Ocean Ridges (Fig. 23-1)

A. Form where two plates are moving away from each other

1. New oceanic lithosphere is formed—basalt, gabbro, periodotite

B. Underwater mountain chain that extends for ~43,000 miles along the earth

C. Characterized by earthquakes and high heat flow

D. Hydrothermal vents

1. Seawater moves through fractures in the crust and dissolves minerals

2. Water is expelled at hot springs

3. When hot water hits cold water of ocean floor, minerals precipitate to form "chimneys"

4. Several hundred species of marine organisms have been discovered

5. Chemosynthesis

a. Bacteria utilize sulfur from the hot springs to make food

b. These bacteria form the base of the food chain at vents

IV. Seafloor Sediment

A. Terrigenous sediment

1. Derived from weathering of the continents

2. Coarser sediment deposited closer to shoreline and finer material away from shoreline

3. Pelagic clay—clay deposits derived from weathering of continents and volcanic islands

B. Biogenous sediment

1. Calcareous oozes

a. Composed of the calcareous ($CaCO_3$) microscopic hard parts of marine plants and animals

b. Chalks form by lithification of calcareous oozes

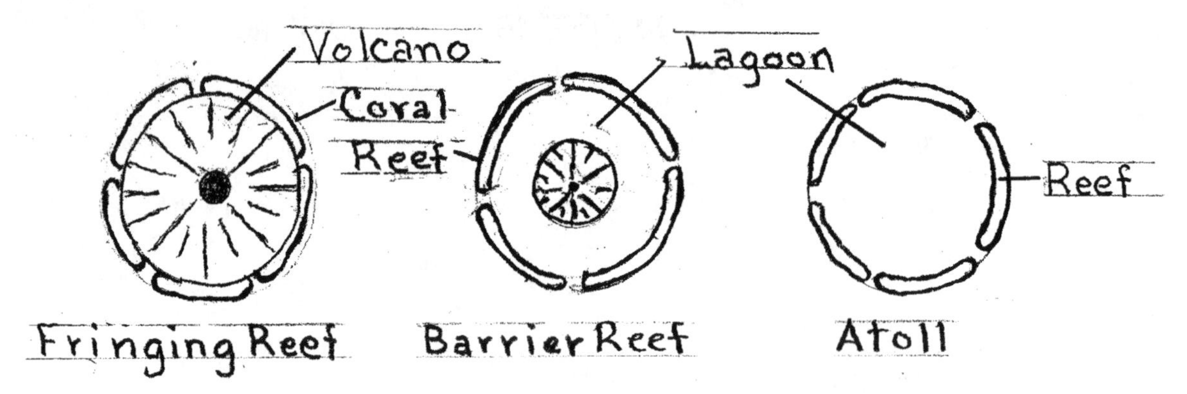


**Figure 23-2** Fringing Reef, Barrier Reef, Atoll
*© A. Troell, 2008*

2. Siliceous oozes
    a. Composed of the siliceous ($SiO_2$) microscopic hard parts of marine plants and animals
    b. Cherts form by lithification of siliceous oozes

V. The Origin of Atolls (Fig. 23-2)
  A. Reefs
    1. Wave-resistant structures built by organisms such as coral and algae
    2. Reef-building corals require warm, shallow, clear marine water
  B. Fringing reefs
    1. Reefs that grow around a volcanic island
  C. Barrier reefs
    1. Reefs that are separated from the volcanic island by a lagoon
  D. Atoll
    1. Rings of coral islands
  E. After Charles Darwin studied atolls in the Pacific Ocean, he hypothesized that the atolls had started out as fringing reefs
    1. As the volcanic island slowly sank and the plate moved, the reefs grew upward and barrier reefs formed
    2. Eventually, the volcanic island sank completely below sea level, but the corals continued to grow upward to form atolls

# NOTES

# NOTES

# CHAPTER 24

## Shorelines

I. Waves

 A. Wave characteristics (Fig. 24-1)

  1. Crest

  2. Trough

  3. Wave height

  4. Wave length

  5. Wave period–time interval between the passage of successive wave crests

  6. Wave base–depth of water motion beneath a passing wave

 B. Waves of oscillation–in the open ocean water particles move in circles

 C. Waves of translation–as wave approaches the shore, water becomes shallower and wave encounters wave base

  1. Wave slows and grows higher, oversteepens, and breaks

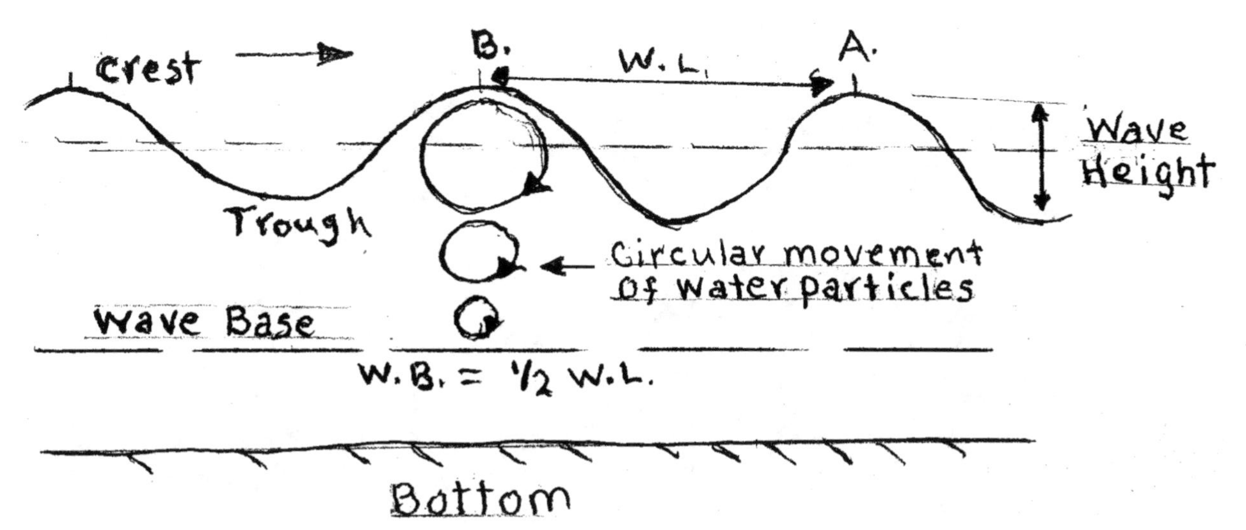


**Figure 24-1** Circular Movement of Water Particles, Wave Height/Base, etc.
*© A. Troell, 2008*

II. Beaches and Shoreline Processes

A. Wave erosion

1. Hydraulic action–impact of wave forces air into cracks, when wave recedes, the air expands and rocks fracture

2. Abrasion–sawing and grinding action of water armed with rock fragments

B. Longshore transport–most waves approach the shore at a slight angle (Fig. 24-2)

1. Beach drift (longshore drift)–zigzag pattern of sand movement along the beach

2. Longshore currents–currents that flow parallel to the shore within the surf zone

C. Wave refraction–bending of waves as they approach (Fig. 24-3)

1. Energy is concentrated against the sides and edges of headlands

2. Over a long period, erosion of headlands and deposition in bays will straighten irregular shorelines

III. Coastal Classification

A. Emergent (erosional) coasts–west coast of North America

1. Develop along active continental margins due to tectonic uplift or subsidence

2. Erosional features (Figs. 24-4 and 24-5)

a. Wave-cut cliffs and wave-cut platforms–created by wave erosion at the base of cliff

b. Wave-cut platforms–flat surface left behind by wave planation

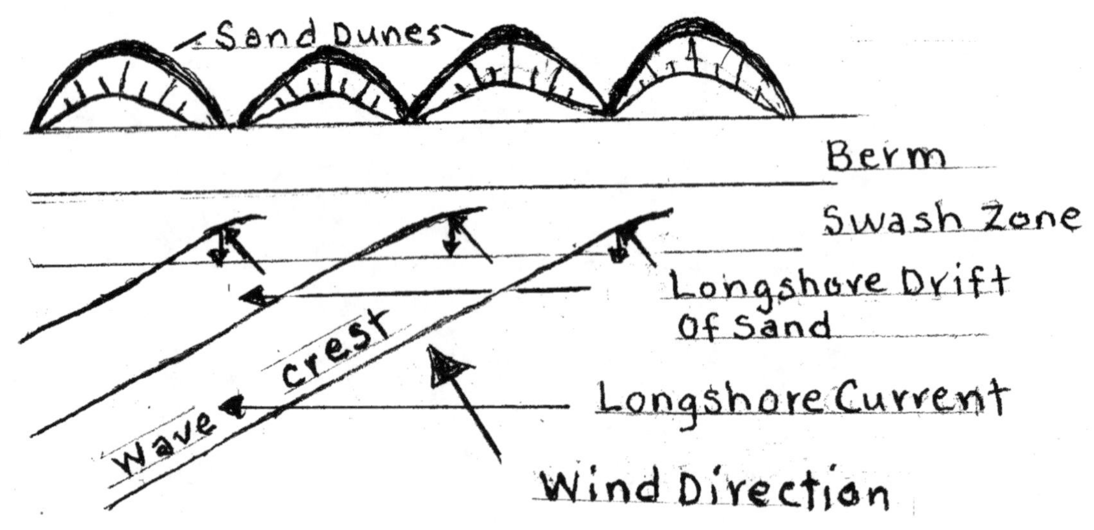


**Figure 24-2** Circular Movement of Water Particles, Wave Height/Base, etc.

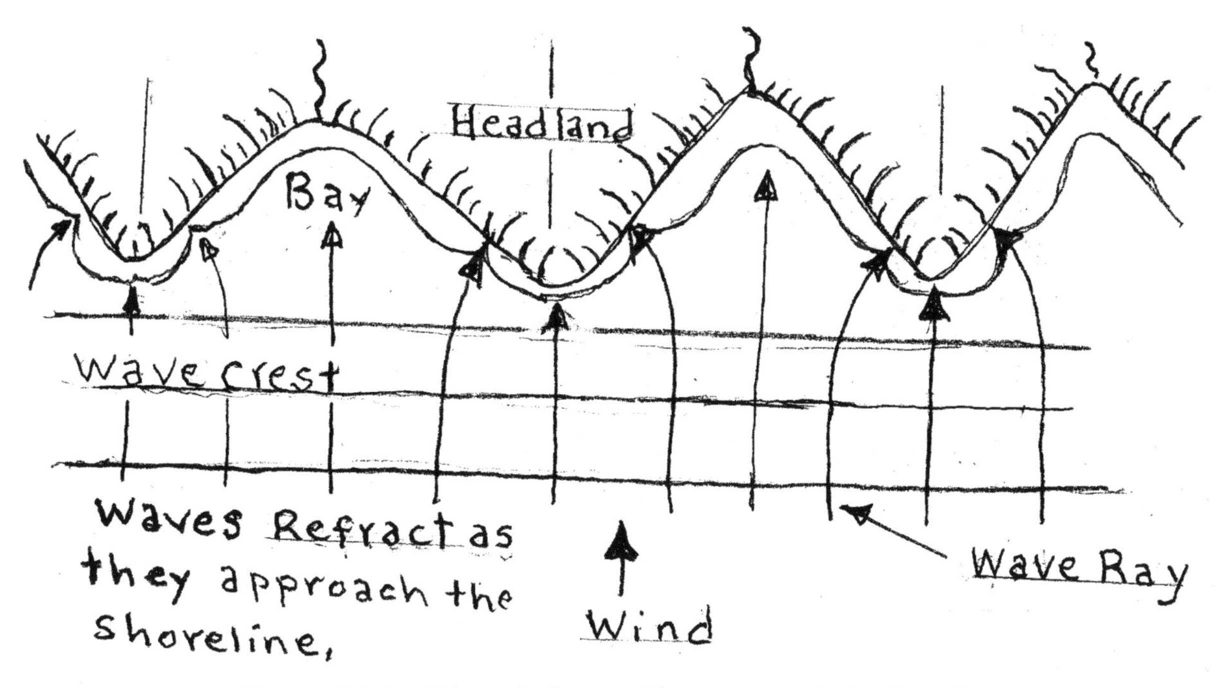


**Figure 24-3** Waves Refract as They Approach the Shoreline
*© A. Troell, 2008*

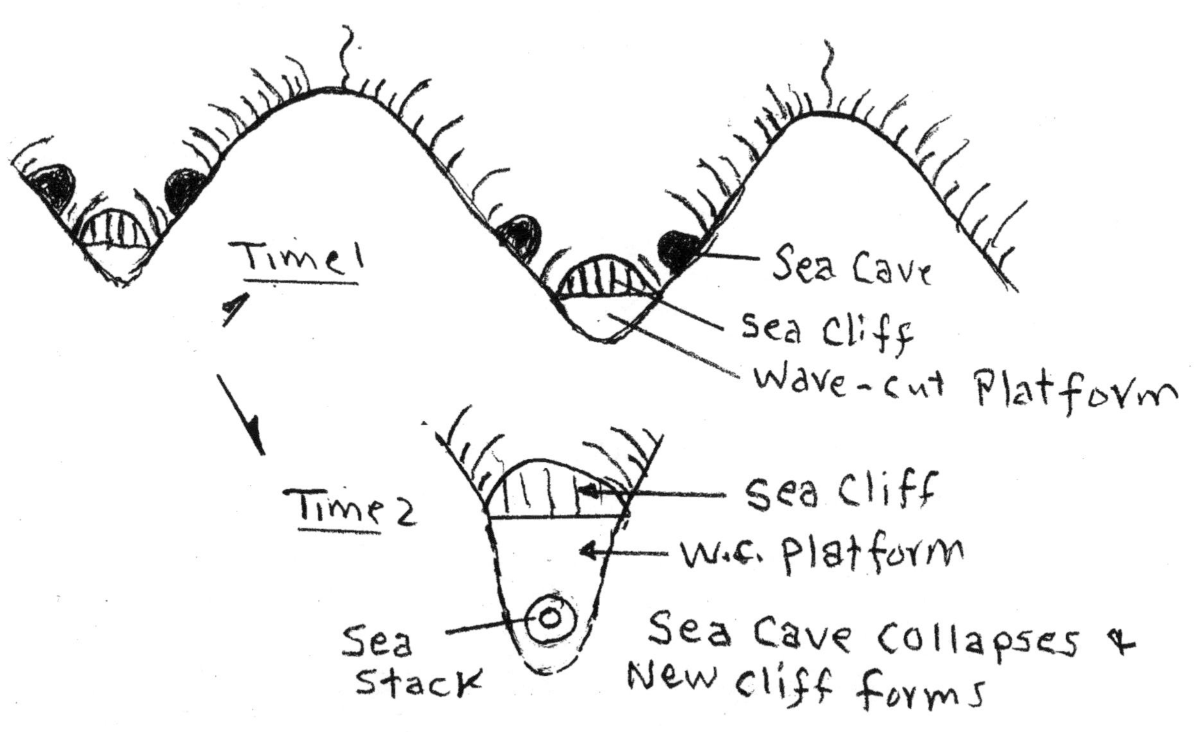


**Figure 24-4** Sea Cave Collapses and New Cliff Forms
*© A. Troell, 2008*

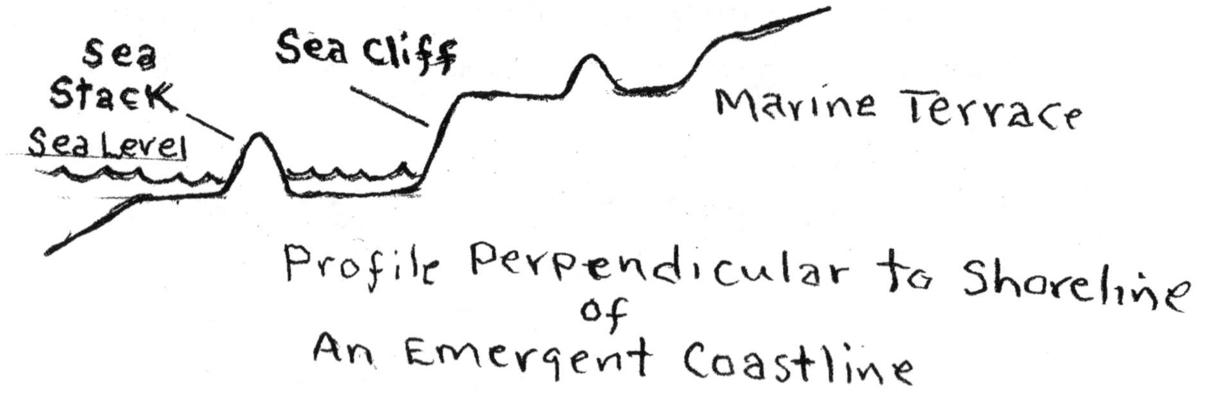


**Figure 24-5** Profit Perpendicular to Shoreline of an Emergent Coastline
*© A. Troell, 2008*

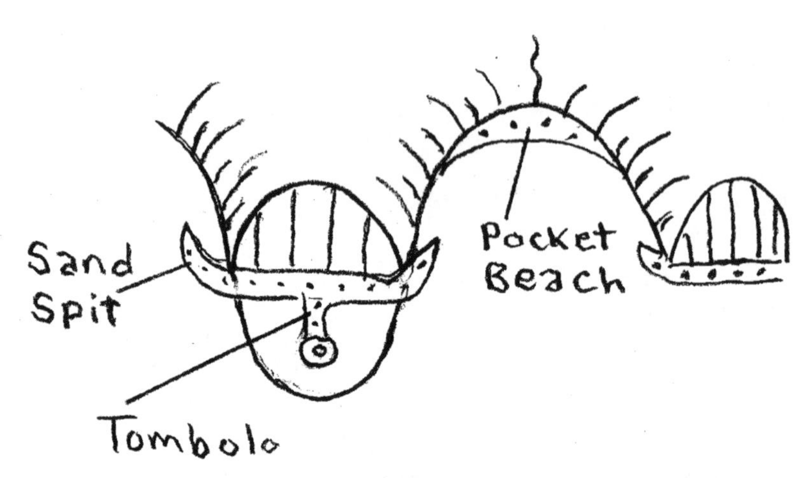


**Figure 24-6** Sand Spit, Pocket Beach, Tombolo
*© A. Troell, 2008*

c. Marine terraces–wave-cut platforms that have been uplifted

d. Sea caves, sea arches, sea stacks

i. Headland erosion produces sea caves

ii. Where two sea caves meet, a sea arch is formed

iii. When the arch falls in, it leaves a sea stack

3. Depositional features (Fig. 24-6)

a. Spits–elongate ridges of sand that extend from land into a bay

b. Pocket beaches between headlands

B. Submergent (depositional) coasts–East coast and Gulf coast of North America

1. Develop along passive continental margins due to rise in sea level or land subsidence

2. Depositional features (Fig. 24-7)

a. Spits

b. Barrier islands

i. Low ridges of sand separated from the shore by a lagoon

ii .Barrier islands extend from New Jersey to the tip of Texas

c. Sand dunes on barrier islands

d. Estuaries–drowned river valleys with muddy floors

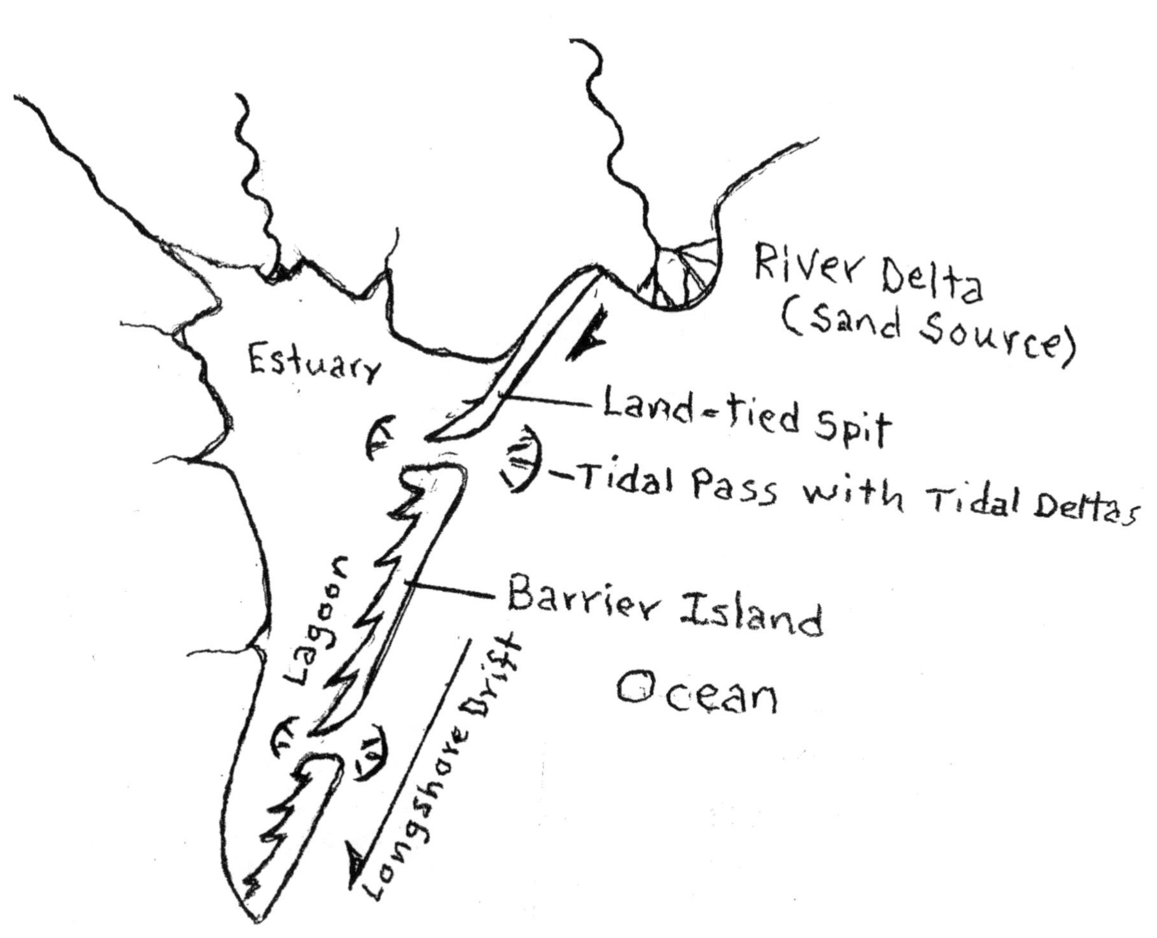


**Figure 24-7** Submergent Coastline

IV. Tides

A. Tides result from the gravitational effects of the moon and, to a lesser extent, the sun (Fig. 24-8)

B. Spring tides

1. Occur during the new and full moons

2. Sun and moon are aligned

C. Neap tides

1. Occur during the first and third quarter moons

2. Sun and moon are at right angles

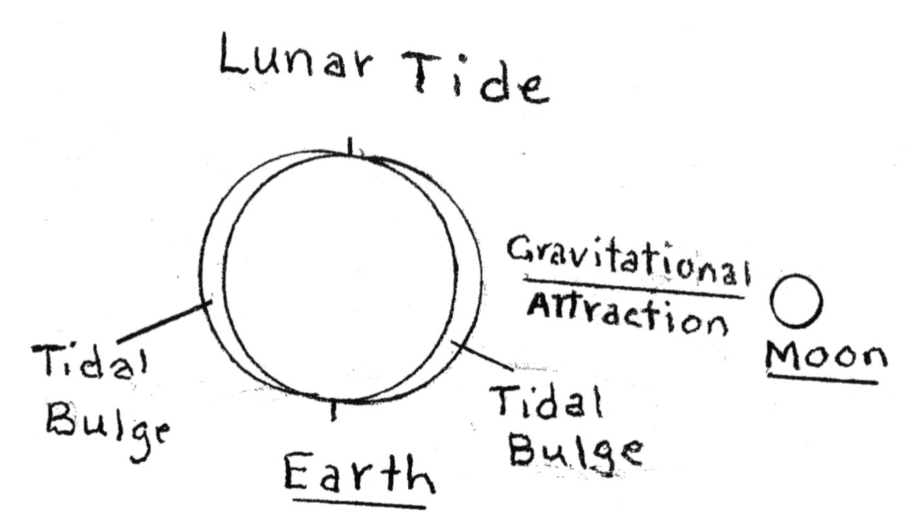


**Figure 24-8** Lunar Tide
*© A. Troell, 2008*

# NOTES

# NOTES

# CHAPTER 25

## The Seasons

I. Earth
   A. Earth rotates on its axis and revolves around the sun
   B. Earth's axis is inclined 23.5 degrees
   C. Axis points in same direction as Earth moves around the sun
   D. Earth's orientation to the sun changes during the year

II. The Seasons (Fig. 25-1)
   A. Because Earth is spherical, only certain areas receive 90-degree rays of the sun
   B. The closer an area is to the area receiving 90-degree rays of the sun
      1. The higher the noon sun angle, the more intense the solar radiation
   C. The 90-degree rays of the sun migrate between 23.5 north of the equator to 23.5 degrees south of the equator during the year
      1. Summer–in Northern Hemisphere, we are closer to the latitude receiving vertical rays of the sun
      2. Winter–in Northern Hemisphere, we are farther away from the latitude receiving vertical rays of the sun
   D. Length of daylight
      1. Summer in Northern Hemisphere–the higher the latitude, the more hours of daylight
      2. Winter in Northern Hemisphere–the higher the latitude, the fewer hours of daylight

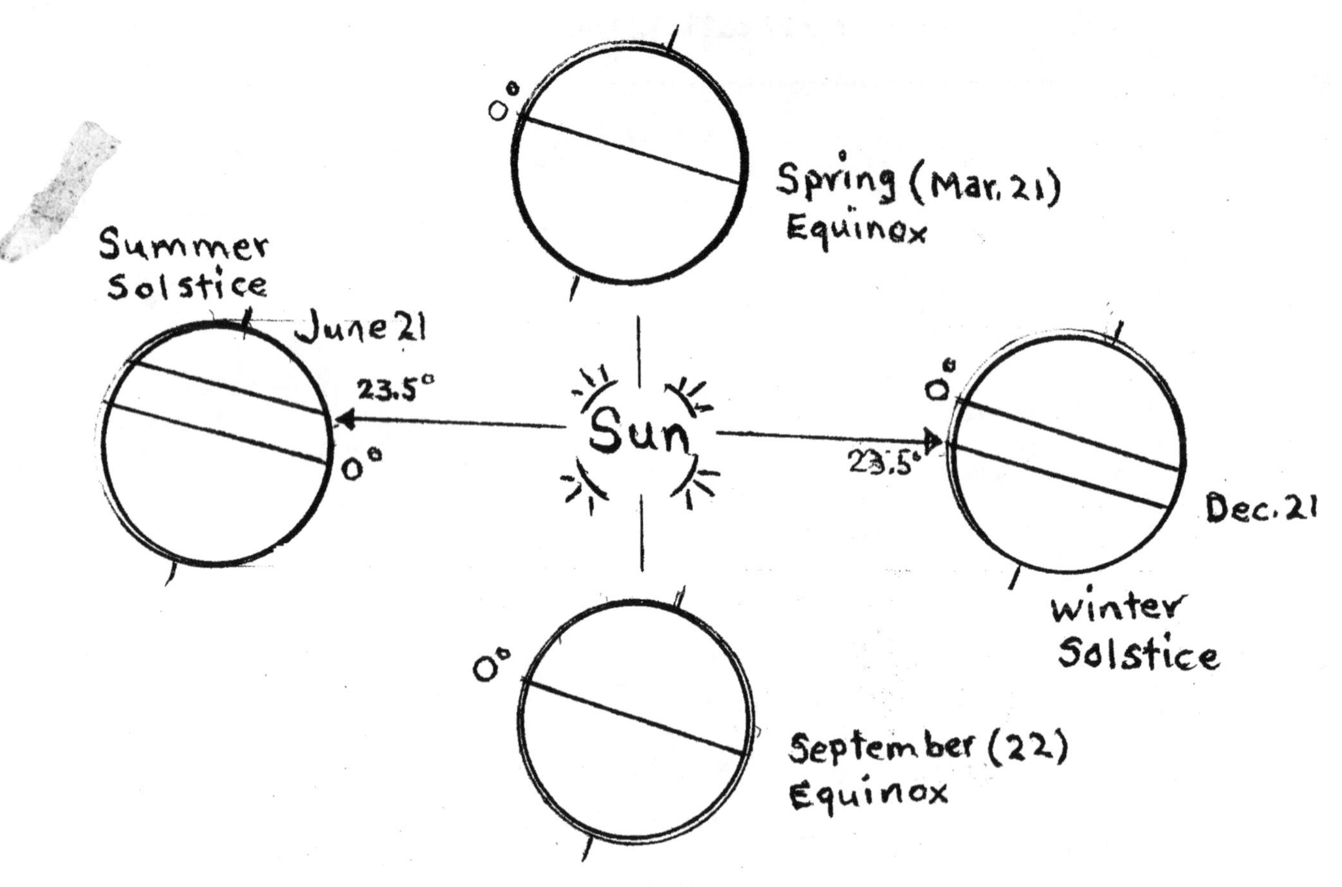


**Figure 25-1** Earth-Sun Relationships
*© A. Troell, 2008*

III. Solstices and Equinoxes

A. Summer Solstice–June 21 or 22

1. Vertical rays of the sun strike the Tropic of Cancer (23½ degrees north of the Equator)

B. Winter Solstice–December 21 or 22

1. Vertical rays of the sun strike the Tropic of Capricorn (23½ degrees south of the Equator)

C. Spring Equinox–March 21 or 22

1. Vertical rays of the sun strike the equator
2. All areas on Earth experience 12 hours of daylight and 12 hours of darkness

D. Autumnal Equinox–September 22 or 23

1. Vertical rays of the sun strike the equator
2. All areas on Earth experience 12 hours of daylight and 12 hours of darkness

# NOTES

# CHAPTER 26

## Clouds

I. Water

   A. Changes of state

      1. Latent heat–the energy absorbed or released during a change of state

      2. Condensation–process whereby water vapor changes to the liquid state; energy is released

      3. Evaporation–liquid water changes to water vapor; energy is absorbed

II. Humidity–Amount of Water Vapor in the Air

   A. Relative humidity

      1. The ratio of the air's actual water vapor content to its potential water vapor capacity at a given temperature

      2. Ways that relative humidity may be changed

         a. Water vapor added to air

         b. Temperature changes

            i. The warmer the air, the more water vapor air can hold

            ii. The cooler the air, the less water vapor it can hold

   B. Dew Point–the temperature to which air would have to be cooled to reach saturation

III. Adiabatic Temperature Changes

   A. Temperature changes that result when air is compressed (and warms) or expands (and cools)

IV. Cloud Formation

   A. In order for clouds to form, air often has to be forced aloft

   B. Ways that air may be forced aloft

      1. Frontal wedging–cold air moves under warm air and forces it upward or warm air moves up over cooler air

2. Orographic lifting–humid air meets mountains and is forced to rise
3. Convective lifting–areas on the ground heat up and air rises
4. Convergence
    a. Air moving horizontally from opposite directions
    b. Florida–air coming from Gulf of Mexico and Atlantic Ocean collide and rise

C. As air rises, it expands and cools

D. As air cools, it can hold less water vapor

E. Eventually air reaches dew point (is saturated) and excess water vapor condenses to form clouds

F. As water vapor condenses, latent heat is released
    1. This keeps air warmer than surrounding air, so that air continues to rise
    2. Release of latent heat fuels severe weather

V. Types of Clouds

A. Air needs to be saturated for clouds to form

B. Clouds are classified by their form and height

C. Basic forms:
    1. Cirrus–high, thin clouds
    2. Cumulus–globular, normally have a flat base
    3. Stratus–sheets or layers that cover much or all of the sky

D. Height:
    1. High clouds–bases are above 18,000–20,000 feet, made up of ice crystals
        a. Cirrus–thin, delicate
        b. Cirrocumulus–high, globular
        c. Cirrostratus–high, thin layers
    2. Middle Clouds (prefix "alto"): 6,500–20,000 feet
        a. Altocumulus–globular masses
        b. Altostratus–uniform white to grayish sheets
    3. Low Clouds: below 6,500 feet
        a. Stratus–foglike layer of clouds
        b. Nimbostratus–layered rain clouds

E. Cumulonimbus

1. Bases may be at low heights, and they may rise to heights of >40,000 feet
2. May produce rain showers or thunderstorms, hail lightning, thunder

VI. Types of Precipitation

A. Rain and snow most common types of precipitation

B. Rain

1. 0.5 mm in diameter
2. Usually associated with nimbostratus or cumulonimbus clouds

C. Drizzle

1. <0.5 mm in diameter
2. Generally produced in stratus or nimbostratus clouds

D. Snow–precipitation in form of ice crystals or aggregates of crystals

E. Sleet

1. Fall of small particles of ice
2. Need warmer layer overlying subfreezing layer

F. Freezing rain (glaze)

1. Subfreezing layer near earth not thick enough to allow raindrops to freeze
2. Raindrops become supercooled (below freezing, but still liquid) and turn to ice when they collide with objects at the earth's surface

G. Hail

1. Precipitation in form of hard, rounded pellets or irregular lumps of ice
2. Produced in large cumulonimbus clouds
3. Ice pellets grow as they collect supercooled water as they fall through the cloud
4. Updrafts may carry hail upward and so hail may acquire additional layers of ice

# NOTES

# CHAPTER 27

## Wind

I. Factors Affecting Winds

   A. Wind–flow of air from areas of high pressure to areas of low pressure

   B. Unequal heating of Earth's surface by the sun is the energy source for most winds

   C. Winds are controlled by

      1. Pressure gradient force

      2. Coriolis effect (Fig. 27-1)

         a. Effect of earth's rotation on moving objects

         b. To right of the target in Northern Hemisphere

         c. To left of the target in Southern Hemisphere

      3. Friction of Earth's surface

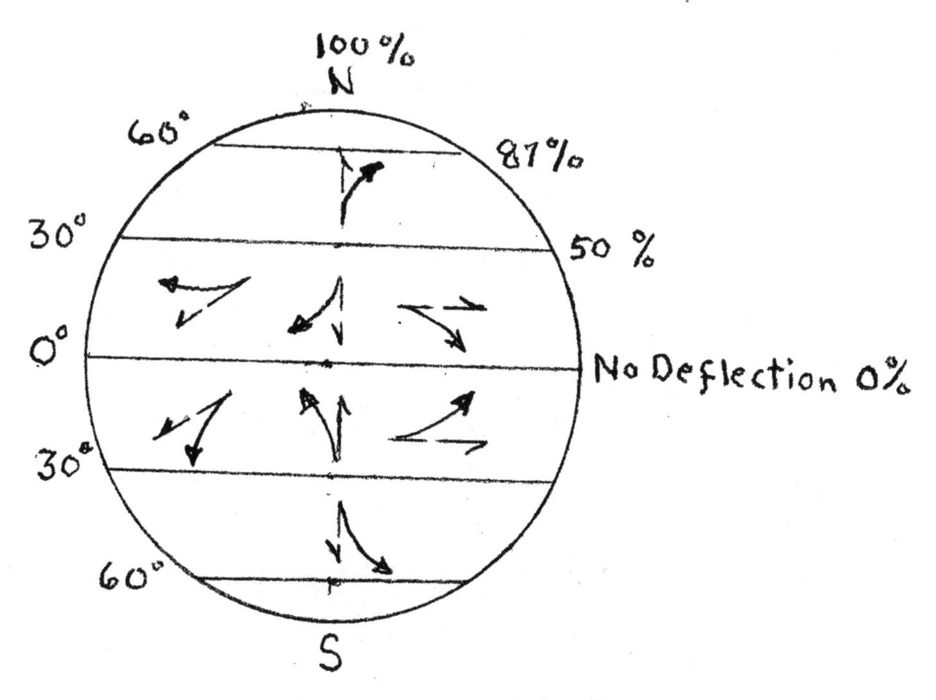


**Figure 27-1** Coriolis Effect
*© A. Troell, 2008*

II. General Circulation of the Earth's Atmosphere (Fig. 27-2)

A. Equatorial low pressure

1. Rising air

2. Abundant precipitation

B. Subtropical high

1. Descending dry air

2. 30 degrees north or south of equator

3. Location of great deserts of Australia, Arabia, and Africa

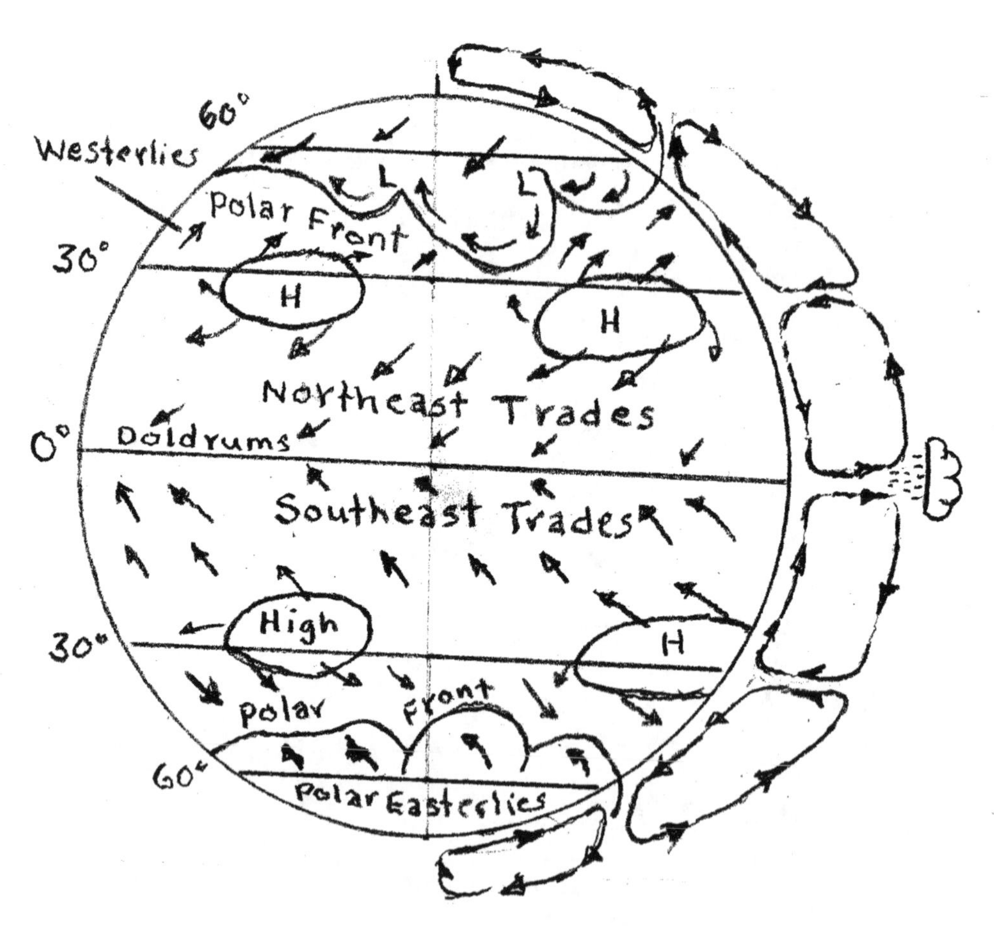


**Figure 27-2** Global Surface Winds
*© After A. N. Strahler*

C. Polar front

1. Results from the interaction of the warm westerly winds and the cold Polar easterly winds
2. Very stormy area

D. Convergence of Trade Winds occurs in a region termed the Intertropical Convergence Zone, which corresponds to the Equatorial Low

# NOTES

# CHAPTER 28

## Air Masses, Fronts, and Severe Weather

I. Air Masses

   A. Large package of air that has similar temperatures and moisture content at any given altitude

   B. Continental polar air mass (cP)–form over land in high latitudes such as Alaska, northern Canada, and Arctic areas

   C. Maritime tropical (mT)–form over warm, moist waters of Gulf of Mexico, Caribbean, or Atlantic

II. Fronts

   A. Boundaries that separate air masses of different temperatures and moisture content

   B. Above Earth's surface, frontal surface slopes at low angle, so warmer air overlies cooler air

   C. Warm fronts (Fig. 28-1)

      1. Warm air moves into an area occupied by cooler air

      2. Usually produce light to moderate precipitation for an extended time

      3. Occasionally associated with thunderstorms when overrunning air is unstable

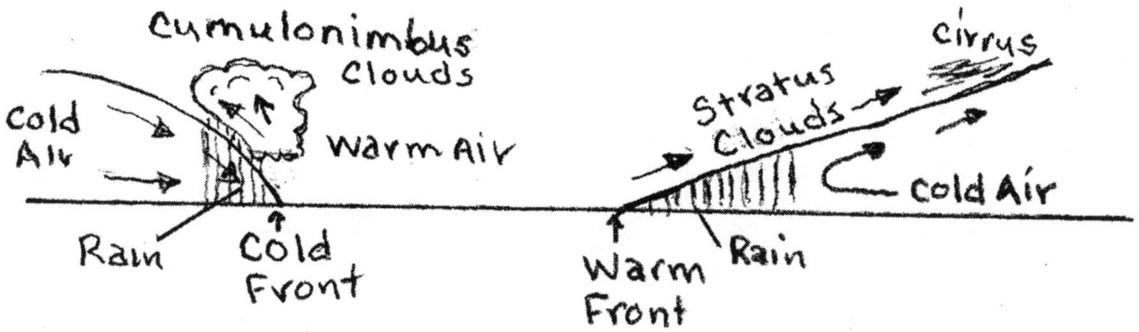

**Figure 28-1** Fronts
*© A. Troell, 2008*

D. Cold fronts (Fig. 28-1)

1. Cooler air moves into an area occupied by warmer air
2. Cold fronts are about twice as steep as warm fronts
3. Move more quickly than warm fronts
4. As cold front approaches, towering clouds are often seen in west or northwest
5. May be associated with thunderstorms and heavy rain

III. Middle-Latitude Cyclone

A. Cyclones (Fig. 28-2)

1. Centers of low pressure that generally travel from west to east
2. Are the primary weather producers in the middle-latitudes

B. Anticyclones—centers of high pressure (Fig. 28-2)

IV. Thunderstorms—Storm that Generates Lightning and Thunder (Fig. 28-3)

A. Stages of thunderstorm development

1. Form when warm, moist air rises in an unstable environment
2. Cumulonimbus clouds grow as warm air rises and releases latent heat
3. Downdrafts develop, releasing heavy precipitation
4. Eventually downdrafts choke off the updrafts and the storm dies

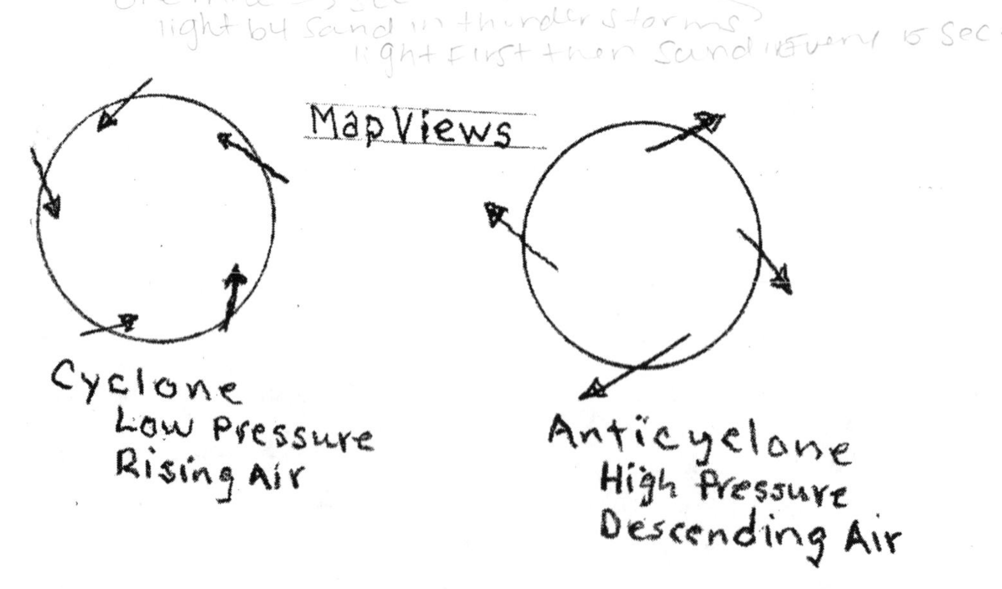


**Figure 28-2** Map Views—Cyclone Low Pressure Rising Air and Anticyclone High Pressure Descending Air
*© A. Troell, 2008*

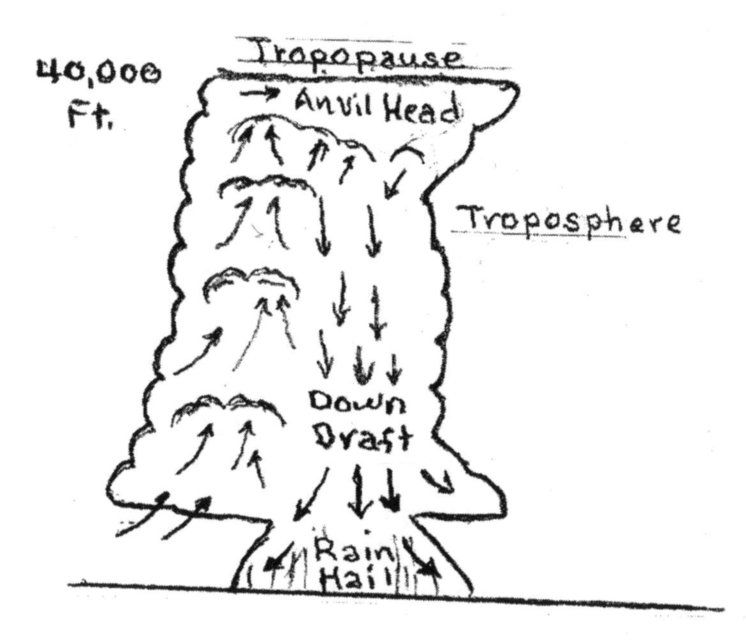


**Figure 28-3** Thunderstorm Cell
*© After A. N. Strahler*

**Figure 28-4** Tornado and Cumulonimbus Cloud
*© After A. N. Strahler*

V. Tornadoes (Fig. 28-4)

A. Violent windstorms that take the form of a rotating column of air that extends downward from a cumulonimbus cloud

B. Mesocyclone

1. Vertical column of rotating air that develops in the updraft of a severe thunderstorm

C. May form along cold front of a middle-latitude cyclone

D. May occur in any month of the year, but April through June is the period of greatest frequency

E. Fujita Intensity Scale–assesses the worst damage produced by storm (F1–F5)

F. Tornado watches–issued when conditions are favorable for tornado formation

G. Tornado warnings–issued when a tornado has been sighted or is indicated by radar

VI. Hurricanes

A. Hurricanes (tropical cyclones)–(Fig. 28-5)

1. Largest storms on Earth
2. Strongest winds near Earth's surface
3. Counterclockwise flow of rising air in Northern Hemisphere

B. Profile of a hurricane

1. Most hurricanes form between 5 and 20 degrees latitude over all tropical oceans except the South Atlantic and eastern South Pacific
2. Form most often in late summer when water temps reach 80 degrees (F)
3. Fueled by release of latent heat

C. Eye–center of the storm where air descends (Fig. 28-6)

D. Eyewall (Fig. 28-7)

1. Doughnut-shaped wall of thunderstorms surrounding the center of the storm
2. Greatest wind speeds and heaviest rainfall occur

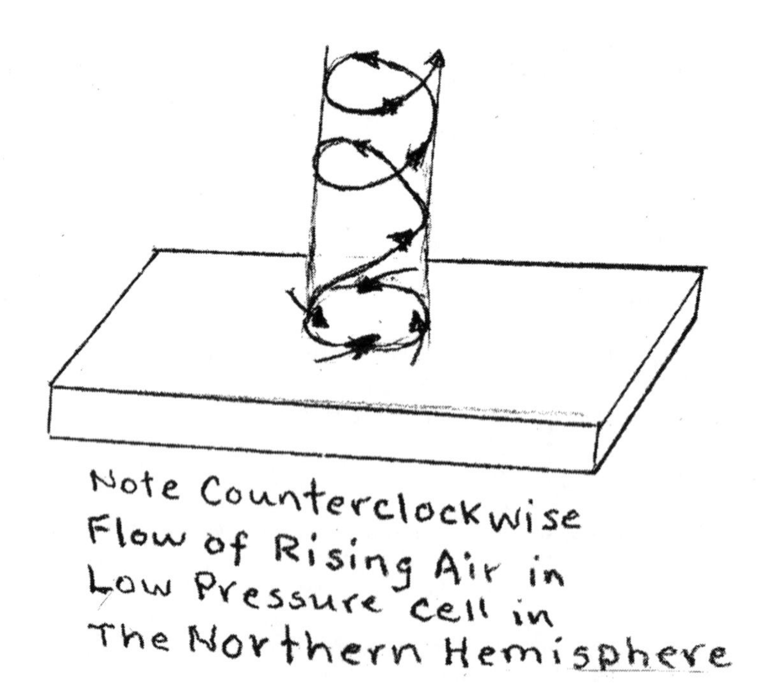


**Figure 28-5** Counterclockwise Flow of Rising Air in Low Pressure Cell in the Northern Hemisphere

**Figure 28-6** Sketch Map of a Hurricane
*© A. Troell, 2008*

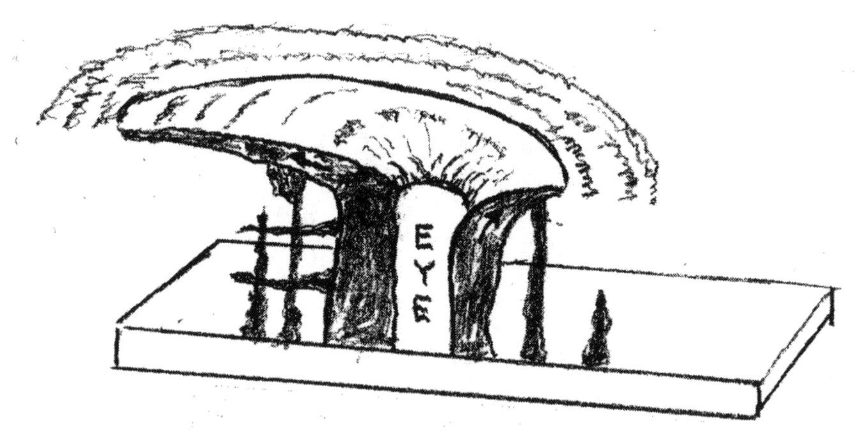


**Figure 28-7** Eye-Wall of a Hurricane
*© After A. N. Strahler*

E. Hurricane formation and decay
   1. Tropical depression (winds do not exceed 38 mph)
   2. Tropical storm (winds are between 38 and 74 mph)
   3. Hurricane (winds exceed 74 mph)
F. Saffir-Simpson scale–ranks the relative intensity of hurricanes (Category 1–5)
G. Hurricane destruction
   1. Storm surge
      a. Dome of water that rises to the right of the eye along a coastline
      b. Responsible for largest percentage of hurricane-related deaths
   2. Wind damage
   3. Inland flooding–heavy rainfall may cause flooding hundreds of miles inland

# NOTES

# CHAPTER 29

## Solar System

I. The First Astronomer to Use the Telescope was Galileo

II. Inner or Terrestrial Planets

   A. Mercury, Venus, Earth, and Mars

   B. Are smaller, denser, and have slower rates of rotation than the Jovian planets

   C. The closer a planet is to the sun, the faster the rate of revolution

      1. Mercury–88 days to orbit the sun

III. Outer or Jovian Planets

   A. Jupiter, Saturn, Uranus, Neptune

   B. Contain a large percentage of gases and have lower densities than the terrestrial planets

   C. All of the outer planets have some type of ring system

IV. Earth's Moon

   A. Craters

      1. Produced by impact of meteoroids

   B. Highlands

      1. Lighter-colored, higher areas

   C. Maria

      1. Dark, smooth areas created by basaltic lava flows

   D. Regolith

      1. Layer of loose sediment produced by bombardment of the surface by micrometeorites

V. Mercury

   A. Cratered highlands and maria like the moon

VI. Venus

A. Similar to Earth in size, density, mass, and location

B. Thick atmosphere

VII. Mars

A. The red planet–color due to iron oxide on the surface

B. Largest known volcano and canyon in the solar system

VIII. Jupiter

A. Largest planet in the solar system

B. Rotates more rapidly than any other planet

C. Has the Great Red Spot

1. Huge rotating storm first observed by Galileo

D. 63 known moons

1. Io is volcanically active

IX. Saturn

A. Large cyclonic storms

B. 60 known moons

X. Uranus

A. Rotates on its side

B. 27 known moons

XI. Neptune

A. 13 known moons

1. Triton is volcanically active

XII. Dwarf Planets

A. Pluto

1. Thin atmosphere

2. One moon called Charon

XIII. Asteroids

A. Most asteroids orbit sun in asteroid belt between Mars and Jupiter

B. Composed of material left over from formation of solar system

XIV. Comets

A. Nucleus

1. Composed of rocky materials and frozen gases

B. Coma

1. As comet nears sun, ices turn to gases

C. Tail(s)

1. Forms as material is blown away from coma

2. Always point away from sun

XV. Meteoroids

A. Most meteor shows consist of material lost by comets

B. Meteoroids that are large enough to survive the trip through Earth's atmosphere are thought to be pieces of asteroids

XVI. Milky Way Galaxy

A. Our solar system is part of the Milky Way galaxy

B. The Milky Way is a large spiral galaxy with at least three distinct spiral arms

C. The sun is located in one of the arms about two-thirds of the way from the center

D. The sun and the arm it is in require about 200 million years for each orbit around the galactic nucleus

# NOTES